IEE Energy Series 3
Series Editors: M. Barak and
                Professor D. T. Swift-Hook

combined
HEAT & POWER
generating systems

combined
HEAT & POWER
generating systems

J. Marecki

Peter Peregrinus Ltd. on behalf of the Institution of Electrical Engineers

Published by: Peter Peregrinus Ltd., London, United Kingdom

British Library Cataloguing in Publication Data

Marecki, Jacek
   Combined heat and power generating systems.
   (IEE energy series; 3).
   1. Heating from central stations
   2. Electric power-plants   3. Waste heat
   I. Title   II. Institution of Electrical
Engineers   III. Series
697'.54   TH7641

ISBN 0 86341 113 4

Printed in England by Short Run Press Ltd., Exeter

Contents

Preface

The aim of this book is to present the main technical and economic aspects of the co-generation of heat and electrical energy in combined heat and power plants. In the first part, the steam cycles applied in combined systems are discussed and the efficiencies of generating heat and electricity in industrial power plants and in combined district heating plants are determined. The characteristics and performance of back-pressure and extraction-condensing turbine sets are described in analytical and graphical forms and several numerical examples of the detailed calculations of power and energy produced in combined systems are given.

In the second part, the economic effects of combined heat and power generation are presented in the form of fuel savings in combined systems compared with the separate systems consisting of equivalent condensing power plants and of heating plants. In this respect, some factors determining the profitability limits of combined generation and the optimisation problems in co-generation plants are discussed. Detailed methods of dividing combined production costs into heat and electrical energy are proposed. Finally, combined gas-steam systems applied in heat and power plants are described.

This book is based on the lectures on power plants and energy economy I have given at the Technical University of Gdańsk for many years and on the results of my study and research works prepared for the national and international conferences in the field of power engineering. The most important references are papers and reports presented to the World Energy Conference and to the International Conference on Industrial Energetics.

To avoid any problems which might arise from the particular technical and economic conditions, existing in this country, all numerical values are given in generally known international units or in the form of relative quantities and the costs are expressed in abstract monetary units. It should, however, be pointed out that the conclusions and detailed results obtained may depend on the specific assumptions and local conditions which can be regarded as examples only.

I am grateful to all those who have helped me in the course of preparing the manuscript, but particularly, I wish to acknowledge the encouragement received from Professor D. T. Swift-Hook of the Institution of Electrical Engineers who asked me to write a book on combined heat and power for his Energy Series, as well as the most valuable help received from Mrs B. Przybylska in preparing this book in English.

Jacek Marecki
Gdańsk, December 1987

List of symbols

a_b – coefficient increasing the investment costs of steam boilers
a_t – coefficient increasing the investment costs of steam turbine sets
B – instantaneous (hourly) fuel consumption
B_{ch} – fuel consumption in a combined system
B_e – fuel consumption for electrical power generation
B_{eb} – fuel consumption for electrical power generation in a back-pressure cycle
B_{ec} – fuel consumption for electrical power generation in a condensing cycle
B_h – fuel consumption for heat output production
B_{sb} – fuel consumption in a steam boiler
B_{sbr} – rated fuel consumption in a steam boiler
B_{se} – fuel consumption in a separate system
B_{wb} – fuel consumption in a hot water boiler
B_{wbr} – rated fuel consumption in a hot water boiler
b – specific fuel consumption
b_b – specific fuel consumption in a balancing power plant
b_{En} – specific fuel consumption for additional net electrical energy
b_{eb} – specific fuel consumption for electrical energy generation in a back-pressure cycle
b_{ec} – specific fuel consumption for electrical energy generation in a condensing cycle
C_W – specific heat of water
c_b – index of specific investment costs for steam boilers
c_{Db} – increment of steam demand by a back-pressure turbine set
c_{De} – increment of steam demand by an extraction-condensing turbine set
c_f – unit price of fuel or fuel heat content
c_{lf} – unit price of liquid fuel
c_{SB} – increment of heat demand by a steam boiler
c_{sf} – unit price of solid fuel
c_T – increment of heat demand by a steam turbine set
c_{Tb} – increment of heat demand by a back-pressure turbine set

c_{Te} – increment of heat demand for electrical power generation in a steam turbine set

c_{Th} – increment of heat demand for heat output production in a steam turbine set

c_t – index of specific investment costs for steam turbine sets

c_u – coefficient depending on the number of units

c_{WB} – increment of heat demand by a hot water boiler

c_w – index of specific investment costs for hot water boilers

D – rate of steam or condensate (water) flow

D_{ad} – make-up water flow compensating condensate losses

D_b – steam flow at the back-pressure turbine inlet/outlet

D_{br} – rated steam flow at the back-pressure turbine inlet/outlet

D_c – steam flow at the condenser inlet

D_e – extraction steam flow

D_{ib} – steam demand by an idle-running back-pressure turbine set

D_{ie} – steam demand by an idle-running extraction-condensing turbine set

D_{rc} – returning condensate flow

D_{rsh} – steam flow in a resuperheater

D_{sb} – capacity of a steam boiler

D_{sbr} – rated capacity of a steam boiler

D_T – steam flow at the turbine inlet

D_0 – steam flow at the boiler outlet

D_1 – steam flow at the first bleeding point

d_b – specific demand for steam by a back-pressure turbine set

d_T – specific demand for steam by a turbine set

E – electrical energy

E_b – electrical energy generated in a back-pressure cycle

E_c – electrical energy generated in a condensing cycle

E_d – daily electrical energy

E_G – electrical energy generated by the gas turbine set in a gas-steam cycle

E_n – net electrical energy

E_S – electrical energy generated by the steam turbine set in a gas-steam cycle

e – exergy

e_b – exergy of steam at the back-pressure turbine outlet

e_{Eb} – exergy of electrical energy generated in a back-pressure cycle

e_{Hb} – exergy of heat delivered in the form of back-pressure steam

e_{sc} – exergy of heat supplied by boilers to the steam cycle

e_0 – exergy of steam at the turbine inlet

F – annual fuel consumption

F_{ch} – annual fuel consumption in a combined system

F_e – annual fuel consumption for electrical energy generation

F_{eb} – annual fuel consumption for electrical energy generation in a back-pressure cycle

F_{ec} – annual fuel consumption for electrical energy generation in a condensing cycle

F_h – annual fuel consumption for heat energy production

F_{sb} – annual fuel consumption in a steam boiler

F_{se} – annual fuel consumption in a separate system

F_{wb} – annual fuel consumption in a hot water boiler

G_w – rate of district heating water flow

G_{wp} – peak-load rate of district heating water flow

H – heat energy

H_a – annual discounted mean value of heat energy

H_b – heat energy delivered by a back-pressure (or extraction) turbine

H_d – daily heat energy

H_{d0} – discounted value of heat energy for the zero year

H_f – calorific value of equivalent fuel

H_G – heat energy delivered in co-operation with the gas turbine set in a gas-steam cycle

H_g – gross heat energy

H_h – heat energy delivered for heating purposes

H_n – net heat energy

H_q – heat energy delivered to the receivers

H_r – heat energy delivered by a reducing valve or a peak-load boiler

H_S – heat energy delivered by the steam turbine set in a gas-steam cycle

H_T – heat energy supplied to a steam turbine set

H_{Tb} – heat energy supplied to a steam turbine set operating in a back-pressure cycle

H_{Tc} – heat energy supplied to a steam turbine set operating in a condensing cycle

H_{Te} – heat energy supplied to a steam turbine set for electrical energy generation

H_{Th} – heat energy supplied to a steam turbine set for heat energy production

H_t – heat energy delivered for process purposes

H_{wb} – heat energy delivered by a hot water boiler

h – enthalpy of steam or water

h_{ad} – enthalpy of make-up water compensating condensate losses

h_{amb} – ambient enthalpy

h_b – enthalpy of steam at the back-pressure turbine outlet

h_{bs} – theoretical (isentropic) value of steam enthalpy at the back-pressure turbine outlet

h_{bt} – theoretical (isentropic) value of steam enthalpy at the back-pressure turbine outlet taking into account the efficiency of steam pipelines

h_c – enthalpy of steam at the condenser inlet

h_{cs} – theoretical (isentropic) value of steam enthalpy at the condenser inlet

h_d – enthalpy of condensate at the condenser outlet

h_e – enthalpy of steam at the extraction point

h_{es} – theoretical (isentropic) value of steam enthalpy at the extraction point
h_{fw} – enthalpy of feed water
h_q – enthalpy of condensate produced from back-pressure (or extraction) steam
h_{rc} – enthalpy of returning condensate
h_{r1} – enthalpy of steam at the resuperheater inlet
h_{r2} – enthalpy of steam at the resuperheater outlet
h_w – resulting enthalpy of condensate and make-up water
h_0 – enthalpy of live steam
h_{0b} – enthalpy of live steam at the boiler outlet
h_{0t} – enthalpy of live steam at the turbine inlet
Δh – drop in steam enthalpy
Δh_b – drop in steam enthalpy in a back-pressure turbine
Δh_{bs} – theoretical (isentropic) drop in steam enthalpy in a back-pressure turbine
Δh_{bt} – theoretical (isentropic) drop in steam enthalpy in a back-pressure turbine taking into account the efficiency of steam pipelines
Δh_c – drop in steam enthalpy in the low-pressure part of an extraction-condensing turbine
Δh_{cs} – theoretical (isentropic) drop in steam enthalpy in the low-pressure part of an extraction-condensing turbine
Δh_d – drop in steam enthalpy in a condenser
Δh_e – drop in steam enthalpy in the high-pressure part of an extraction-condensing turbine
Δh_{es} – theoretical (isentropic) drop in steam enthalpy in the high-pressure part of an extraction-condensing turbine
Δh_q – drop in steam enthalpy in a back-pressure steam receiver
Δh_{qs} – theoretical (isentropic) drop in steam enthalpy in a back-pressure steam receiver
Δh_0 – drop in steam enthalpy in live steam pipelines
I – investment costs (outlays)
I_{sb} – partial investment costs falling to the steam boilers
I_t – partial investment costs falling to the turbogenerator sets
I_{wb} – partial investment costs falling to the hot water boilers
K – annual costs
K_b – annual costs of generating energy in a back-pressure cycle
K_c – annual fixed costs
K_{cE} – annual fixed costs falling to electrical energy
K_{cH} – annual fixed costs falling to heat energy
K_E – annual costs falling to electrical energy
K_{Eb} – annual value of electrical energy generated in a back-pressure cycle
K_f – annual fuel costs
K_H – annual costs falling to heat energy
K_i – investment costs

K_{id} – sum of investment costs discounted to the zero year
K_o – annual operating costs
K_{oa} – annual discounted mean value of operating costs
K_{oc} – annual fixed operating costs
K_{od} – sum of operating costs discounted to the zero year
K_{ov} – annual variable operating costs
K_r – annual reproduction costs
K_{tr} – annual costs of heat transport
K_v – annual variable costs
K_{vE} – annual variable costs falling to electrical energy
K_{vH} – annual variable costs falling to heat energy
ΔK – difference in annual costs (in favour of a combined system)
ΔK_c – difference in annual fixed costs
ΔK_E – difference in annual costs apportioned to electrical energy
ΔK_H – difference in annual costs apportioned to heat energy
ΔK_v – difference in annual variable costs
k – specific cost
k_c – specific fixed cost
k_{cE} – specific fixed cost of electrical energy
k_{cH} – specific fixed cost of heat energy
k_E – specific cost of electrical energy
k_{Eb} – unit value of electrical energy generated in a back-pressure cycle
k_H – specific cost of heat energy
k_{hp} – specific investment cost in a heating plant
k_i – specific investment cost
k_{qT} – specific cost of heat supplied to the steam turbine
k_{sb} – specific investment cost for the steam boilers
k_t – specific investment cost for the turbogenerator sets
k_{tr} – specific cost of heat transport
k_v – specific variable cost
k_{vE} – specific variable cost of electrical energy
k_{vH} – specific variable cost of heat energy
k_{wb} – specific investment cost for the hot water boilers
k_z – coefficient of 'freezing' the investment costs
L – length of a heat transmission pipeline
m_a – annual load factor
m_{ae} – annual electrical load factor
m_{ah} – annual heat load factor
m_b – load factor in the back-pressure cycle
m_d – daily load factor
m_{de} – daily electrical load factor
m_{dh} – daily heat load factor
N – number of operation years
N_{sb} – number of steam boilers in a heat and power plant

N_t – number of turbogenerator sets in a heat and power plant

N_{wb} – number of hot water boilers in a heat and power plant

n – number of stages in the feed water heating scheme

n_{pp} – ratio of the peak-load electrical power to the rated electrical power

n_{qp} – ratio of the peak-load heat output to the rated heat output

P – electrical or mechanical power

P_{av} – average (mean) value of the electrical power

P_b – electrical power generated in a back-pressure cycle

P_{be} – electrical power supplied to the power system by a back-pressure turbogenerator set

P_{bn} – net electrical power of a back-pressure turbogenerator set

P_{bp} – peak-load electrical power generated in a back-pressure cycle

P_{br} – rated electrical power of a back-pressure turbogenerator set

P_c – electrical power generated in a condensing cycle

P_{cp} – peak-load electrical power generated in a condensing cycle

P_e – electrical power supplied to the power system by a turbogenerator set

P_i – internal power output of a steam turbine

P_{ib} – internal power output generated in a back-pressure cycle

P_{ic} – internal power output generated in a condensing cycle

P_{ih} – internal power output generated in the high-pressure part of a steam turbine

P_{il} – internal power output generated in the low-pressure part of a steam turbine

P_m – mechanical power output on the turbine shaft

P_{mb} – mechanical power output on the back-pressure turbine shaft

P_{min} – minimum value of the electrical power

P_n – net electrical power of a turbogenerator set

P_p – peak-load value of the electrical power

P_T – gross electrical power of a turbogenerator set

P_{Tr} – rated electrical power of a turbogenerator set

P_ε – auxiliary electrical power demand

$P_{\varepsilon e}$ – auxiliary power demand connected with electrical power generation

$P_{\varepsilon h}$ – auxiliary power demand connected with heat production

ΔP – electrical power losses

p – accumulation rate

p_a – share of electrical energy generated in a back-pressure cycle in the annual electrical energy production

p_b – steam pressure at the back-pressure turbine outlet (back-pressure)

p_c – steam pressure at the condenser inlet

p_e – steam pressure at the extraction valve of an extraction-condensing turbine (extraction pressure)

p_p – share of electrical power generated in a back-pressure cycle in the peak-load electrical power

p_0 – live steam pressure

p_{0b} – live steam pressure at the boiler outlet
p_{0t} – live steam pressure at the turbine inlet
Δp – drop in steam pressure
Q – heat flow (heat demand, heat output)
Q_{av} – average (mean) value of the heat output
Q_b – heat output of a back-pressure (or extraction) turbine
Q_{bmax}– maximum heat output of a back-pressure (or extraction) turbine
Q_{bp} – peak-load heat output of a back-pressure (or extraction) turbine
Q_{br} – rated heat output of a back-pressure (or extraction) turbine
Q_G – heat output delivered in co-operation with the gas turbine set in a gas-steam cycle
Q_g – gross heat output
Q_h – heat output delivered for heating purposes
Q_{hp} – peak-load heat demand for heating purposes
Q_i – heat demand of an idle-running turbine set
Q_{ib} – heat demand of an extraction-condensing turbine set when idle-running in a back-pressure cycle
Q_{ic} – heat demand of an extraction-condensing turbine set when idle-running in a condensing cycle
Q_{ie} – electromechanical component of the heat demand of an idle-running turbine set
Q_{ih} – heat component of the heat demand of an idle-running turbine set
Q_n – net heat output
Q_p – peak-load heat output
Q_q – heat output delivered for room heating
Q_{qp} – peak-load heat demand for room heating
Q_r – heat output of a reducing valve or peak-load boiler
Q_{rp} – peak-load heat output of a reducing valve or peak-load boiler
Q_S – heat output delivered by the steam turbine set in a gas-steam cycle
Q_{SB} – heat flow supplied to a steam boiler
Q_{sb} – heat output delivered by a steam boiler
Q_T – heat flow supplied to a steam turbine set
Q_{Tb} – heat flow supplied to a steam turbine set operating in a back-pressure cycle
Q_{Tc} – heat flow supplied to a steam turbine set operating in a condensing cycle
Q_{Te} – heat flow supplied to a steam turbine set for electrical power generation
Q_{Th} – heat flow supplied to a steam turbine set for heat output production
Q_{Tr} – rated heat flow supplied to a steam turbine
Q_t – heat output delivered for process purposes
Q_{tp} – peak-load heat demand for process purposes
Q_v – heat output delivered for ventilation purposes
Q_{vp} – peak-load heat demand for ventilation purposes

Q_{WB} – heat flow supplied to a hot water boiler
Q_{wb} – heat output delivered by a hot water boiler
Q_{wbp} – peak-load heat output of a hot water boiler
Q_{wbr} – rated heat output of a hot water boiler
Q_0 – heat flow supplied to a heat and power plant
Q_{0e} – heat flow supplied to a heat and power plant for electrical power generation
Q_{0h} – heat flow supplied to a heat and power plant for heat output production
ΔQ – heat losses
q – specific heat demand
q_b – specific heat demand of a back-pressure turbine set
q_{be} – specific heat demand of a back-pressure turbine set for electrical power generation
q_c – specific heat demand of a steam turbine for electrical power generation in a condensing cycle
q_i – specific heat demand of an idle-running turbine set
q_{ie} – electromechanical component of the specific heat demand of an idle-running turbine set
q_{ih} – heat component of the specific heat demand of an idle-running turbine set
q_q – specific heat demand for room heating
q_{qd} – specific heat demand for room heating in dwelling houses
q_{qi} – specific heat demand for room heating in industrial buildings
q_{qp} – specific heat demand for room heating in public utility buildings
q_T – specific heat demand of a steam turbine set
q_v – specific heat demand for ventilation
q_{vd} – specific heat demand for ventilation in dwelling houses
q_{vi} – specific heat demand for ventilation in industrial buildings
q_{vp} – specific heat demand for ventilation in public utility buildings
R – fuel equivalent of fixed costs
r – annual reproduction rate
r_c – coefficient of annual fixed costs
r_o – coefficient of annual fixed operating costs
s – entropy
s_a – annual value of heat/electricity ratio (energy index)
s_{amb} – ambient entropy
s_p – peak-load value of heat/power ratio (energy index)
T – period of a year (8760 hours)
T_{amb} – ambient temperature
T_{av} – average (mean) temperature
T_{av0} – average temperature at which heat is supplied to a plant
T_{avq} – average temperature at which heat is delivered by a plant

T_{bep} – utilisation time of the peak-load electrical power of a back-pressure turbine set

T_{ber} – utilisation time of the rated electrical power of a back-pressure turbine set

T_{bhp} – utilisation time of the peak-load heat output of a back-pressure turbine set

T_{bhr} – utilisation time of the rated heat output of a back-pressure turbine set

T_{bp} – utilisation time of the peak-load back-pressure output

T_{br} – utilisation time of the rated back-pressure output

T_{cp} – utilisation time of the peak-load condensing power

T_d – period of a day (24 hours)

T_{di} – daily utilisation time of the installed capacity

T_{dp} – daily utilisation time of the peak-load

T_{dpe} – daily utilisation time of the peak-load electrical power

T_{dph} – daily utilisation time of the peak-load heat output

T_h – duration time of the heating load

T_{hp} – utilisation time of the peak-load heat output for heating purposes

T_p – utilisation time of the peak-load

T_{pe} – utilisation time of the peak-load electrical power

T_{ph} – utilisation time of the peak-load heat output

T_r – duration time of the heat output of a reducing valve or peak-load boiler

T_{rh} – utilisation time of the rated heat output

T_{rp} – utilisation time of the peak-load heat output of a reducing valve or peak-load boiler

T_t – duration time of the process heat load

T_{tp} – utilisation time of the peak-load heat output for process purposes

T_v – duration time of the rated external temperature for ventilation purposes

T_w – operation (working) time

T_{wp} – utilisation time of the peak-load heat output of hot water boilers

t – real time

t_b – steam temperature at the back-pressure turbine outlet

t_d – load duration time

t_e – steam temperature at the extraction valve

t_{es} – steam saturation temperature at the extraction valve

t_s – steam saturation temperature

t_0 – live steam temperature

t_{0b} – live steam temperature at the boiler outlet

t_{0t} – live steam temperature at the turbine inlet

Δt – drop in steam temperature

u_{ba} – share of heat energy delivered by a back-pressure (or extraction) turbine in the annual heat energy production

u_{bp} – share of heat output delivered by a back-pressure (or extraction) turbine in the peak-load heat output

u_G – share of heat output delivered in co-operation with the gas turbine set in a gas-steam cycle

u_{Ga} – share of annual heat energy delivered in co-operation with the gas turbine set in a gas-steam cycle

u_{ha} – share of heat energy delivered for heating purposes in the annual heat energy production

u_{hp} – share of heat output delivered for heating purposes in the peak-load heat output

u_{ra} – share of heat energy delivered by the reducing valve or peak-load boiler in the annual heat energy production

u_{rp} – share of heat output delivered by the reducing valve or peak-load boiler in the peak-load heat output

u_S – share of heat output delivered by the steam turbine set in a gas-steam cycle

u_{Sa} – share of annual heat energy delivered by the steam turbine set in a gas-steam cycle

u_{ta} – share of heat energy delivered for process purposes in the annual heat energy production

u_{tp} – share of heat output delivered for process purposes in the peak-load heat output

V – cubature of buildings

V_d – cubature of dwelling houses

V_i – cubature of industrial buildings

V_p – cubature of public utility buildings

w_b – exponential related to the investment costs of steam boilers

w_t – exponential related to the investment costs of turbogenerator sets

w_w – exponential related to the investment costs of hot water boilers

x_{ce} – share of the fixed costs falling to electrical energy in the annual fixed costs of a heat and power plant

x_{ch} – share of the fixed costs falling to heat energy in the annual fixed costs of a heat and power plant

x_e – share of costs falling to electrical energy in the annual costs of a heat and power plant

x_h – share of costs falling to heat energy in the annual costs of a heat and power plant

x_{ib} – relative value of the heat demand of an idle-running turbine set in a back-pressure cycle

x_{ih} – relative value of the heat component of idle-running losses

x_{ve} – share of the variable costs falling to electrical energy in the annual variable costs of a heat and power plant

x_{vh} – share of the variable costs falling to heat energy in the annual variable costs of a heat and power plant

y_e – coefficient of the non-utilisation of extraction steam

z – coefficient of flow mixing in a heat receiving system

α – combined base-load factor

α_c – share of condensing steam flow in the total steam flow of a turbine

α_e – share of extraction steam flow in the total steam flow of a turbine

α_p – combined base-load factor at the peak-load time

β – coefficient correcting the idle-running losses of an extraction-condensing turbine set

γ – ratio of the liquid fuel price to the solid fuel price

δ – coefficient increasing the heat demand of a back-pressure turbine set

δ_c – coefficient decreasing the fixed costs in a heat and power plant

δ_E – conversion coefficient for electrical energy generated in a substitute condensing power plant

δ_H – conversion coefficient for heat energy delivered by a substitute heating plant

δ_P – conversion coefficient for electrical power generated in a substitute condensing power plant

δ_Q – conversion coefficient for heat output delivered by a substitute heating plant

δv – coefficient decreasing the variable costs in a heat and power plant

ε – share of auxiliary power demand

ε_e – share of auxiliary power demand connected with electrical power generation

ε_q – coefficient of the exergetic value of heat delivered by outlet steam

ε_S – share of auxiliary power demand in the steam part of a gas-steam power plant

ε_0 – coefficient of the exergetic value of heat supplied to inlet steam

$\Delta\varepsilon$ – drop in the share of auxiliary power demand

η – efficiency

η_{au} – efficiency of the auxiliaries

η_b – efficiency of a boiler

η_{bc} – efficiency of a back-pressure cycle

η_c – efficiency of a steam cycle

η_{cc} – efficiency of a condensing cycle

η_{ce} – efficiency of an extraction-condensing cycle

η_e – partial efficiency of the electrical energy generation

η_{eb} – partial efficiency of the electrical energy generation in a back-pressure cycle

η_{ec} – partial efficiency of the electrical energy generation in a condensing cycle

η_{em} – electromechanical efficiency of a turbogenerator set

η_{ex} – exergetic efficiency

η_{fw} – efficiency of a feed-water pump

η_G – Gašparović's efficiency of a heat and power plant

η_g – efficiency of a generator
η_h – partial efficiency of the heat energy production
η_{he} – efficiency of a heat exchanger
η_i – internal efficiency of a steam turbine
η_{ih} – internal efficiency of the high-pressure part of a steam turbine
η_{il} – internal efficiency of the low-pressure part of a steam turbine
η_m – mechanical efficiency of a turbogenerator set
η_{me} – efficiency of an electrical motor
η_n – net efficiency
η_{ne} – partial net efficiency of electrical energy generation
η_{nh} – partial net efficiency of heat energy production
η_{pi} – efficiency of steam pipelines
η_{sb} – efficiency of a steam boiler
η_t – efficiency of a transmission gear
η_{tr} – efficiency of a transformer
$\eta_{tr\varepsilon}$ – efficiency of an auxiliary transformer
η_{wb} – efficiency of a hot water boiler
η_ε – coefficient taking into account the auxiliary power demand
μ – relative operating time of a turbogenerator set
ϱ – index of electrical power generation in a back-pressure cycle
σ_a – annual combined production index
σ_g – gross combined production index
σ_n – net combined production index
σ_p – peak-load combined production index
σ_r – rated combined production index
$\Delta\sigma$ – increment of the combined production index
τ_{be} – water temperature after base-load heat exchanger
τ_{ex} – external temperature
τ_{exr} – rated external temperature
τ_{exv} – rated external temperature for ventilation
τ_{in} – internal temperature
τ_q – water temperature at the inlet of inside heating installations
τ_{qp} – peak-load temperature at the inlet of inside heating installations
τ_{w1} – water temperature at the hot water boiler inlet
τ_{w2} – water temperature at the hot water boiler outlet
τ_1 – outlet temperature of district heating water
τ_{1p} – peak-load value of the outlet temperature of district heating water
τ_2 – return temperature of district heating water
τ_{2p} – peak-load value of the return temperature of district heating water
$\Delta\tau$ – temperature difference in the heat exchanger
φ_c – coefficient correcting the division of fixed costs in a heat and power plant
φ_{ce} – coefficient correcting the division of fixed costs in a heat and power plant with extraction-condensing turbogenerator sets

φ_r – regeneration coefficient taking into account the feed-water heating

φ_v – coefficient correcting the division of variable costs in a heat and power plant

φ_{ve} – coefficient correcting the division of variable costs in a heat and power plant with extraction-condensing turbogenerator sets

BP   – combined heat and power plant with back-pressure turbogenerator sets

CH   – combined heat and power plant

CP   – condensing power plant

EP   – combined heat and power plant with extraction-condensing turbogenerator sets

HN   – heat network

HP   – heating plant

TL   – power transmission line

Chapter 1

Theory of combined heat and power

1.1 Steam cycles in heat and power plants

Combined heat and power systems involve the simultaneous cogeneration of electrical and heat energy in the form of low-pressure steam or hot water. The power installations enabling the combined production of both types of energy form a combined system in which the required energy conversions take place.

The conversion of chemical energy contained in fuel into the heat energy of combustion gases and then that of steam takes place in high-pressure steam boilers. Further conversion of this energy into mechanical and electrical energy, and the heat energy of low-pressure steam, takes place in turbogenerator sets consisting of steam turbines and synchronous generators. Hot water is obtained from heat exchangers supplied with low-pressure steam or separate water boilers.

Energy conversion taking place in a combined system is presented in the form of a flow diagram in Fig. 1.1, where the following cycles of the main energy carriers of the system are shown:

- fuel cycle: from storage yard (1), through coal-handling plant (2), boiler combustion chamber (3) and ash remover (4), to slag and ash dump (5);
- air and exhaust gas cycle: from air ventilator (6), through boiler heating surface (7) and draught ventilator (8), to the chimney;
- main steam-water cycle: from feed pump (9), through boiler heating surfaces (10), (11), and (12), steam turbine (13), condenser (14) and condensate pump (15), to feed-water tank (17);
- cooling-water cycle with condenser (14) and cooling-water pump (18) – open, closed or mixed.

Combined systems are installed in heat and power stations. The type of heat and power station depends on the type of steam turbines, which may be back-pressure, back-pressure with extraction or extraction-condensing. Heat and power plants can also be divided into *district heating* plants which supply the heat distribution network with heat energy mainly for heating purposes and *industrial* which supply process heat mainly for technological purposes.

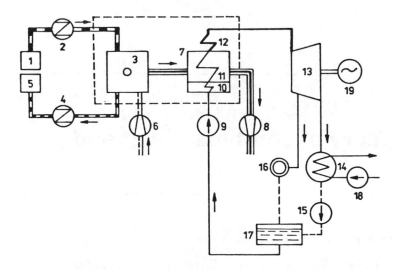

Fig. 1.1   *Block diagram of energy conversions in a combined system 1 – fuel storage, 2 – coal-handling plant, 3 – combustion chamber, 4 – ash-handling plant, 5 – ash-storage yard, 6 – air ventilator, 7 – boiler heating surfaces, 8 – draught fan, 9 – feed-water pump, 10 – economiser, 11 – steam boiler, 12 – steam superheater, 13 – steam turbine, 14 – condenser, 15 – condensate pump, 16 – heat receiver, 17 – feed-water tank, 18 – cooling-water pump, 19 – generator*

The steam cycle comprising back-pressure turbines in a heat and power plant is called a back-pressure cycle. Fig. 1.2 shows this cycle in the T, s diagram and Fig. 1.3 the same cycle in the h, s diagram. Fig. 1.4 gives a simplified heat diagram of a back-pressure heat and power plant.

The theoretical steam cycle shown in both entropic diagrams consists of the following processes:

– isobaric preheating 1–2 in the gas-water economiser;
– isobaric (isothermal) evaporation 2–3 in the boiler evaporator;
– isobaric superheating 3–4 in the steam superheater;
– adiabatic expansion of steam 4–5 in the turbine, constituting an irreversible process;
– isobaric cooling of steam 5–6 to saturation state;
– isobaric (isothermal) condensation of steam 6–7–8 in the steam receiver;
– isentropic pumping of water 8–1 in the pump, it being assumed in Figs. 1.2 and 1.3 that points 8 and 1 coincide due to the very small increase in the water enthalpy in the pump.

The first three processes 1–2–3–4 take place in the boiler with an inlet pressure p_0 and processes 5–6–7–8 in the steam receiver with outlet pressure (back-pressure) p_b. Steam pressure losses in the superheater and steam pipelines, also the overcooling of condensate from the outlet steam, that is to say the lowering of its temperature in relation to the saturation temperature,

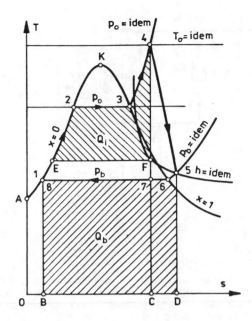

Fig. 1.2 *Back-pressure cycle in the T, s diagram*

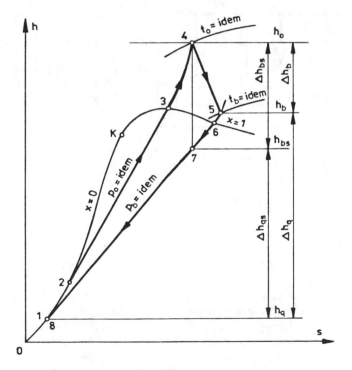

Fig. 1.3 *Back-pressure cycle in the h, s diagram*

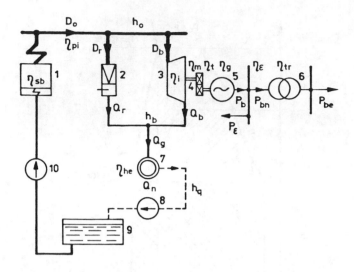

Fig. 1.4 *Heat diagram of a back-pressure heat and power plant 1 – steam boiler, 2 – steam reducing valve, 3 – steam turbine, 4 – transmission gear, 5 – generator, 6 – transformer, 7 – heat receiver, 8 – condensate pump, 9 – feed-water tank, 10 – feed-water pump*

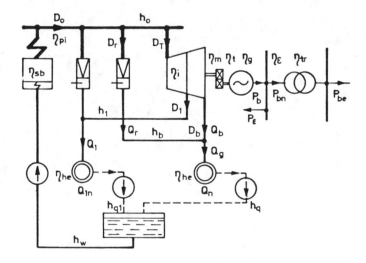

Fig. 1.5 *Heat diagram of an extraction-backpressure heat and power plant*

taking place in the real cycle, have not been taken into account. In the theoretical back-pressure cycle (without loss in the turbine), the expansion of steam is along the line 4–7 instead of 4–5. The real irreversible process then substitutes the reversible adiabatic curve (isentrope). The heat converted into mechanical work in the theoretical cycle is presented in the *T, s* diagram by field 1–2–3–4–7–8–1 and the heat contained in the low-pressure outlet steam by field B–8–7–C–B. The theoretical drops in enthalpy in the turbine $\Delta h_{bs} = h_4 - h_7$ and in the steam receiver $\Delta h_{qs} = h_7 - h_8$ are presented in the h, s diagram.

A simplified heat diagram of a heat and power station with extraction-backpressure turbines is presented in Fig. 1.5. A turbine with one bleeding point gives a steam flow D_1 with a pressure p_1 and a flow D_b with a pressure p_b from the back-pressure outlet. If there are several bleeding points, the bleeding steam pressures amount to p_1, p_2 etc. Steam from the bleeding points is delivered to receivers, which require a suitably high pressure as opposed to steam delivered from regeneration bleeding points, which serves to feed the regenerating feed-water heaters.

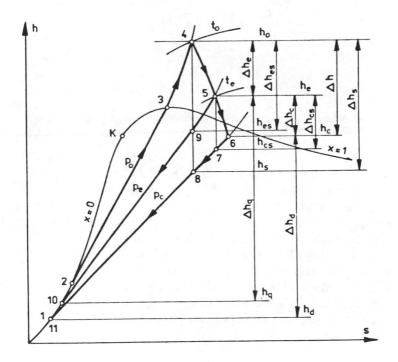

Fig. 1.6   *Extraction-condensing cycle in the h, s diagram*

The extraction-condensing cycle is presented in the h, s diagram, in Fig. 1.6 and a simplified heat diagram of a heat and power station with an extraction-condensing turbine in Fig. 1.7. The extraction steam pressure is p_e and the condenser pressure is p_c. Corresponding steam flows are shown in the diagram.

Heat converted into mechanical energy now consists of two parts, namely, heat supplied to the turbine in steam flow D_T, for which the enthalpy drop is Δh_e, and heat supplied by the flow of steam condensed in condenser D_c, for which the enthalpy drop is Δh_c. Heat supplied to the extraction steam

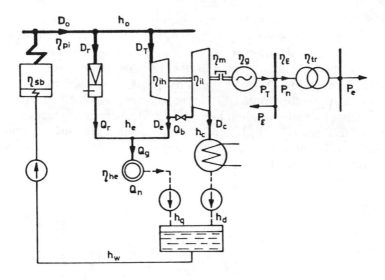

Fig. 1.7   *Heat diagram of an extraction-condensing heat and power plant with a reduction-cooling valve*

receivers is proportional to the enthalpy drop Δh_q and heat supplied to the cooling water corresponds to the enthalpy drop Δh_d.

The essential difference between extraction-backpressure and extraction-condensing cycles is that in the former, the extraction steam flow D_1 changes depending on the back-pressure steam flow D_b and does not usually exceed several per cent of this flow. On the other hand, in the extraction-condensing turbine both steam flows D_e and D_c are, for all practical purposes, independent, as the turbine can work both in a solely condensing cycle wih $D_e = 0$ and in an almost back-pressure cycle with a flow $D_c = D_{c\,min}$, required in view of the operating conditions of the turbine outlet and the condensing installations.

1.2 Power and energy generated in combined systems

1.2.1 Power and energy in a back-pressure system

The gross electrical power generated at the generator terminals in a back-pressure system and the gross heat output at the back-pressure outlet can be calculated from the energy balance of the turbine set which is presented schematically in Fig. 1.4. This balance is determined by the equations

$$P_b = D_b \, \Delta h_b \, \eta_{em} \qquad (1.1)$$

$$Q_b = D_b \, \Delta h_q \qquad (1.2)$$

here

$$\Delta h_b = h_0 - h_b = (h_0 - h_{bs})\, \eta_i = \Delta h_{bs}\, \eta_i \qquad (1.3)$$

$$\Delta h_q = h_b - h_q = h_0 - \Delta h_{bs}\, \eta_i - h_q \qquad (1.4)$$

$$\eta_{em} = \eta_m\, \eta_t\, \eta_g \qquad (1.5)$$

where:
P_b – electrical power output generated in a back-pressure cycle, kW;
D_b – rate of flow of steam through the turbine, kg/s;
Q_b – back-pressure heat output (gross), kJ/s;
Δh_b – real drop in enthalpy in the turbine, kJ/kg;
Δh_{bs} – theoretical (isentropic) drop in enthalpy in the turbine;
Δh_q – real drop in enthalpy in the steam receiver;
h_0 – enthalpy of steam at the turbine inlet;
h_b – enthalpy of steam at the turbine outlet;
h_{bs} – theoretical value of outlet enthalpy in an isentropic process;
h_q – enthalpy of condensate produced from output steam;
η_i – internal efficiency of turbine;
η_{em} – electromechanical efficiency of turbogenerator set;
η_m – mechanical efficiency;
η_t – efficiency of transmission gear;
η_g – generator efficiency.

When applying a coherent system of units (SI), quantitative formulae (1.1) and (1.2) do not require any additional calculation coefficients. In other systems of units, coefficients occur in these equations depending upon the units applied, hence corresponding numerical formulae emerge.

In power engineering, the unit of 1 t/h = 1/3·6 kg/s is frequently applied as a legal unit of mass rate of flow and 1 MW = 10^3 kW as a derivative unit of both electrical and heat power. In such a system of units, the following equations are obtained instead of (1.1) and (1.2):

$$3600\, P_b = D_b\, \Delta h_b\, \eta_{em} \qquad (1.6)$$

$$3600\, Q_b = D_b\, \Delta h_q \qquad (1.7)$$

here: P_b – electrical power output generated in a back-pressure cycle, MW; D_b – rate of flow of steam, t/h; Q_b – back-pressure heat output, MJ/s.

In the theoretical, closed back-pressure cycle, which is shown in Figs. 1.2 and 1.3, the loss of condensate is not taken into account and in formula (1.4), the value of the enthalpy of saturated water at pressure p_b is accepted as the enthalpy of the condensate h_q. On the other hand, in the real back-pressure cycle, losses of condensate of up to several score per cent occur. Then instead of the condensate enthalpy h_q the weighted mean of

$$h_w = \frac{D_{rc}\, h_{rc} + D_{ad}\, h_{ad}}{D_{rc} + D_{ad}} \qquad (1.8)$$

is assumed, where

D_{rc} – the flow rate of returning condensate input to the cycle after subtracting losses;

h_{rc} – enthalpy of returning condensate input to the cycle;

D_{ad} – flow of make-up water compensating losses of condensate in cycle;

h_{ad} – enthalpy of make-up water.

Depending on the energy balance, one can calculate the specific steam and heat demand by the back-pressure turbine set referred to the gross power output

$$d_b = \frac{D_b}{P_b} = \frac{1}{\Delta h_b \, \eta_{em}} \tag{1.9}$$

$$q_b = d_b(h_0 - h_q) = \frac{\Delta h_b + \Delta h_q}{\Delta h_b \, \eta_{em}} \tag{1.10}$$

here:

d_b – specific demand for steam;

q_b – specific demand for heat.

Of the above heat demand by the back-pressure turbine set the fraction for electrical energy generated amounts to

$$q_{be} = d_b(h_0 - h_b) = \frac{1}{\eta_{em}} \tag{1.11}$$

The specific heat demand for the generation of back-pressure energy thus depends solely on the electromechanical efficiency of the turbogenerator set η_{em}, and not on the internal efficiency of the turbine η_i. This can also be explained by the h, s diagram (Fig. 1.3), from which it results that the enthalpy difference $h_5 - h_7$, arising as the result of heat losses in the turbine, does not mean heat lost externally, as in the condensing cycle, but corresponds to the increase in heat supplied to the steam receivers.

The difference between the gross power P_b at the generator terminals and the power consumed by auxiliary installations P_ε (Fig. 1.4) is defined as the net electrical power of a back-pressure heat and power station P_{bn}. The power supplied to the power system P_{be} is further diminished by losses in the unit transformer. These relations can be defined by the formula

$$P_{be} = P_{bn} \, \eta_{tr} = P_b \, \eta_\varepsilon \, \eta_{tr} = P_b(1 - \varepsilon) \, \eta_{tr} \tag{1.12}$$

here:

$\varepsilon = P_\varepsilon/P_b$ – the relative consumption of power for auxiliary demands;

η_{tr} – the efficiency of unit transformer.

The electrical power P_ε, consumed for auxiliary demands in a heat and power station, can be divided into two parts:

$P_{\varepsilon e}$ – auxiliary power demand connected with electrical power generation and

$P_{\varepsilon h}$ – auxiliary power demand connected with the generation of heat, which is taken to be an external consumer.

In such case, the value

$$\varepsilon_e = \frac{P_{\varepsilon e}}{P_b}$$

which is slightly lower than ε, is taken to be the relative energy consumption for auxiliary electricity demands.

The gross heat output of the back-pressure heat and power station Q_g is the sum of the outputs from the back-pressure outlet Q_b, and from the reducing-cooling valve Q_r connected in parallel, which serves to cover the short-term heat load and also as a standby installation for the back-pressure turbine. If heat is supplied to consumers through a heat exchanger, most frequently a steam-water heat exchanger, then the net heat output is further diminished by losses in the installation, which results from the relationship

$$Q_n = Q_g \, \eta_{he} = (Q_b + Q_r) \, \eta_{he} \qquad (1.13)$$

where η_{he} – the efficiency of the heat exchanger.

The gross electrical energy generated in a back-pressure system is calculated on the basis of the corresponding power value (peak or rated value) and the utilisation time of this power, from the formula

$$E_b = P_{bp} \, T_{bp} = P_{bp} \, m_b \, T \qquad (1.14)$$

in which:
E_b – gross electrical energy, MWh/a;
P_{bp} – peak-load power output at the back-pressure turbogenerator terminals, MW;
T_{bp} – annual utilisation time of peak-load back-pressure output, h/a;
m_b – load factor in back-pressure cycle;
$T = 8760$ h/a.

The heat energy supplied from the back-pressure outlet is calculated similarly, on the basis of the corresponding value of heat output, where it can be approximately assumed that the utilisation time of this output is equal to the utilisation time of electrical back-pressure power. Thus

$$H_b = Q_{bp} \, T_{bp} \qquad (1.15)$$

where:
H_b – heat energy delivered from back-pressure outlet, MJ/a;
Q_{bp} – peak-load heat output from this outlet, MJ/s;
T_{bp} – annual utilisation time of peak-load back-pressure output, s/a[1].

[1] In accordance with instructions regarding the application of SI units and other legal units of measurement, electrical energy is expressed in kWh, MWh and multiples of kWh, and heat energy – correspondingly in kJ, MJ and further multiples of joules. Electrical power is expressed in kW, MW and further multiples of watts, and heat output is expresed in kJ/s, MJ/s etc. Thus the annual utilisation time of electrical power is defined in hours, as 1 MWh/(MW·a) = 1 h/a, whereas the annual utilisation time of heat output should be defined in seconds, as 1 MJ/(MJ·s^{-1}·a) = 1 s/a. The conversion from seconds to hours can easily be carried out using the multiplier 3600 s/h.

To calculate the total gross heat energy H_g and the net heat energy H_n, supplied by a heat and power station, the energy supplied from the back-pressure outlet H_b and that from the reducing-cooling valve H_r should be added. Thus

$$H_n = H_g \, \eta_{he} = (H_b + H_r) \, \eta_{he} \qquad (1.16)$$

here

$$H_r = Q_{rp} \, T_{rp} \qquad (1.17)$$

where:
Q_{rp} – peak-load heat output of the reducing-cooling valve;
T_{rp} – annual utilisation time of this peak-load output, s/a.

Example 1.1
A back-pressure turbogenerator set delivers a heat output $Q_b = 25$ MJ/s in steam with an outlet pressure $p_b = 0.5$ MPa, and inlet parameters:
(a) 3·4 MPa, 435 °C;
(b) 8·8 MPa, 500 °C.
    In each case the efficiencies of the turbogenerator set are $\eta_i = 0.75$, $\eta_{em} = 0.92$. Calculate outlet steam flow, gross electrical power and specific demand for steam and heat.

Solution:
(a) Assuming inlet parameters of 3·4 MPa, 435 °C from the h, s diagram, we obtain $h_0 = 3303$ kJ/kg; with a back-pressure $p_b = 0.5$ MPa, the theoretical (isentropic) enthalpy value is $h_{bs} = 2809$ kJ/kg; from the saturated steam table $h_q = 636$ kJ/kg.
    The theoretical (isentropic) enthalpy drop in the turbine is

$$\Delta h_{bs} = 3303 - 2809 = 494 \text{ kJ/kg}$$

the real drop from formula (1.3) is

$$\Delta h_b = 494 \cdot 0.75 = 372 \text{ kJ/kg}$$

Hence the steam enthalpy at the turbine outlet is

$$h_b = 3303 - 372 = 2931 \text{ kJ/kg}$$

and the drop in enthalpy in steam receiver

$$\Delta h_q = 2931 - 636 = 2295 \text{ kJ/kg}$$

    Steam flow corresponding to the given heat output is determined from formula (1.2)

$$D_b = \frac{Q_b}{\Delta h_q} = \frac{25000}{2295} = 10.9 \text{ kg/s} = 39.2 \text{ t/h}$$

and the electrical power at the turbogenerator terminals amounts to

$$P_b = D_b \, \Delta h_b \, \eta_{em} = 10 \cdot 9 \cdot 372 \cdot 0 \cdot 92 = 3730 \text{ kW} = 3 \cdot 73 \text{ MW}$$

The specific demand for steam according to formula (1.9) amounts to

$$d_b = \frac{D_b}{P_b} = \frac{39 \cdot 2}{3 \cdot 73} = 10 \cdot 5 \text{ kg/kWh}$$

or

$$d_b = \frac{D_b}{P_b} = \frac{10 \cdot 9}{3730} = 2 \cdot 92 \cdot 10^{-3} \text{ kg/kJ}$$

and the specific demand for heat for the total production of electrical and heat energy, according to formula (1.10)

$$q_b = d_b(h_0 - h_q) = 10 \cdot 5(3303 - 636) \cdot 10^{-3} = 28 \cdot 0 \text{ MJ/kWh}$$

or

$$q_b = 2 \cdot 92 \cdot 10^{-3}(3303 - 636) = 7 \cdot 79 \text{ kJ/kJ}.$$

The specific demand for heat, falling to electrical energy generated by back-pressure steam, according to formula (1.11) is

$$q_{be} = d_b(h_0 - h_b) = 10 \cdot 5(3303 - 2931) \cdot 10^{-3} = 3 \cdot 91 \text{ MJ/kWh}$$

or

$$q_{be} = \frac{1}{\eta_{em}} = \frac{1}{0 \cdot 92} = 1 \cdot 09 \text{ kJ/kJ}$$

(b) Assuming inlet parameters of 8·8 MPa, 500 °C, we correspondingly obtain:

$$h_0 = 3391 \text{ kJ/kg}$$
$$h_{bs} = 2688 \text{ kJ/kg}$$
$$\Delta h_{bs} = 703 \text{ kJ/kg}$$
$$\Delta h_b = 703 \cdot 0 \cdot 75 = 528 \text{ kJ/kg}$$
$$h_b = 3391 - 528 = 2863 \text{ kJ/kg}$$
$$\Delta h_q = 2863 - 636 = 2227 \text{ kJ/kg}$$
$$D_b = \frac{25000}{2227} = 11 \cdot 2 \text{ kg/s} = 40 \cdot 4 \text{ t/h}$$
$$P_b = 11 \cdot 2 \cdot 528 \cdot 0 \cdot 92 \cdot 10^{-3} = 5 \cdot 44 \text{ MW}$$
$$d_b = \frac{40 \cdot 4}{5 \cdot 44} = 7 \cdot 42 \text{ kg/kWh}$$
$$q_b = 7 \cdot 42(3391 - 636) \cdot 10^{-3} = 20 \cdot 4 \text{ MJ/kWh}$$
$$q_{be} = 7 \cdot 42(3391 - 2863) \cdot 10^{-3} = 3 \cdot 91 \text{ MJ/kWh}$$

Example 1.2
Calculate the gross annual electrical and heat energy output by a back-pressure turbogenerator set from Example 1.1 (with a peak-load heat output $Q_{bp} = 25$ MJ/s) and by a reduction-cooling valve with a peak-load heat output $Q_{rp} = 30$ MJ/s working in parallel, if the annual utilisation time of the back-pressure peak-load power output is $T_{bp} = 4000$ h/a, and that of the reduction-cooling valve is $T_{rp} = 1000$ h/a.

Solution:
(a) If the inlet parameters are 3·4 MPa, 435 °C, the gross peak-load electrical power at the turbogenerator terminals amounts to $P_{bp} = 3·73$ MW. According to the formula (1.14), the gross annual electrical energy is

$$E_b = P_{bp} \, T_{bp} = 3·73 \cdot 4000 \cdot 10^{-3} = 14·92 \text{ GWh/a}$$

According to formulae (1.15) and (1.17), the annual heat energy from the back-pressure turbine is

$$H_b = Q_{bp} \, T_{bp} = 25 \cdot 4000 \cdot 3600 \cdot 10^{-6} = 360 \text{ TJ/a}$$

and from the reduction-cooling valve

$$H_r = Q_{rp} \, T_{rp} = 30 \cdot 1000 \cdot 3600 \cdot 10^{-6} = 108 \text{ TJ/a}$$

that is to say that the total gross heat energy amounts to

$$H_g = H_b + H_r = 360 + 108 = 468 \text{ TJ/a}$$

(b) If the inlet parameters are 8·8 MPa, 500 °C, then
$$P_{bp} = 5·44 \text{ MW}$$

$$E_b = P_{bp} \, T_{bp} = 5·44 \cdot 4000 \cdot 10^{-3} = 21·76 \text{ GWh/a}$$

$$H_g = H_b + H_r = 468 \text{ TJ/a}$$

1.2.2 Power and energy in extraction systems
The energy balance of an extraction-backpressure turbogenerator set, shown in Fig. 1.5, is defined by a set of equations

$$P_b = [D_l(h_0 - h_l) + D_b(h_0 - h_b)]\eta_{em} \qquad (1.18)$$

$$Q_l = D_l(h_l - h_{ql}) \qquad (1.19)$$

$$Q_b = D_b(h_b - h_q) \qquad (1.20)$$

here

$$D_T = D_l + D_b \qquad (1.21)$$

where:
P_b – electrical power output generated in a back-pressure cycle;
D_l – extraction steam flow;
D_b – outlet steam flow;

D_T – flow of steam delivered to the turbine;
Q_b – heat output delivered from the back-pressure outlet, whereas the enthalpy values are given in Fig. 1.5.

The specific demand for steam and heat, as well as the net values of electrical and heat energy are calculated as in an exclusively back-pressure cycle.

A simplified heat diagram of an extraction-condensing heat and power plant is presented in Fig. 1.7. In this cycle, the electrical power of the turbogenerator set P_T has two components:

– power obtained in the back-pressure cycle, generated in the high-pressure part of the turbine on the extraction steam flow D_e, is called in brief: back-pressure power P_b;
– power obtained in the condensing cycle, generated in both parts of the turbine on the steam flow condensed in condenser D_c, is called in brief: condensing power P_c.

The internal efficiency of the turbine generally differs in the high-pressure and low-presure parts. Thus the energy balance is defined by a set of equations

$$P_b = D_e \, \Delta h_e \, \eta_{em} \tag{1.22}$$

$$P_c = D_c \, \Delta h \, \eta_{em} \tag{1.23}$$

$$Q_b = D_e \, \Delta h_q \tag{1.24}$$

here

$$P_T = P_b + P_c \tag{1.25}$$

$$D_T = D_e + D_c \tag{1.26}$$

$$\Delta h = h_0 - h_c = \Delta h_e + \Delta h_c \tag{1.27}$$

$$\Delta h_e = h_0 - h_e = (h_0 - h_{es}) \, \eta_{ih} = \Delta h_{es} \, \eta_{ih} \tag{1.28}$$

$$\Delta h_c = h_e - h_c = (h_e - h_{cs}) \, \eta_{il} = \Delta h_{cs} \, \eta_{il} \tag{1.29}$$

$$\Delta h_q = h_e - h_q = h_0 - \Delta h_{es} \, \eta_{ih} - h_q \tag{1.30}$$

where:
P_b – electrical power output generated in a back-pressure cycle;
P_c – electrical power output generated in a condensing cycle;
P_T – total gross electrical power of the turbogenerator set;
D_e – flow of extraction steam;
D_c – flow of condensing steam;
D_T – flow of steam delivered to the turbine;
Q_b – gross heat output of the turbine, delivered from extraction;
η_{ih} – internal efficiency of the high-pressure part;
η_{il} – internal efficiency of the low-pressure part of the turbine.

The enthalpy values and the theoretical and real drops in enthalpy are given in the h, s diagram (Fig. 1.6). The enthalpy values are also determined in the heat diagram (Fig. 1.7).

Formulae serving the calculation of total electrical power or total steam flow delivered to the extraction-condensing turbogenerator set are obtained from the conversion of equations (1.22–1.30)

$$P_T = (D_e \, \Delta h_e + D_c \, \Delta h) \, \eta_{em} \qquad (1.31)$$

or

$$P_T = (D_T \, \Delta h_e + D_c \, \Delta h_c) \, \eta_{em} \qquad (1.32)$$

and

$$D_T = \frac{P_T}{\eta_{em} \, \Delta h} + D_e y_e \qquad (1.33)$$

here y_e in the formula (1.33), called the coefficient of non-utilisation of extraction steam, is defined as the quotient of drops in enthalpy

$$y_e = \frac{\Delta h_c}{\Delta h} = \frac{\Delta h_c}{\Delta h_e + \Delta h_c} \qquad (1.34)$$

Equation (1.33) can also be used to calculate the total flow of steam D_T under condensing operation conditions, when $D_e = 0$ as the result of closing the extraction. On this basis, the specific demand for steam $d_T = D_T/P_T$ can be determined.

The net electrical power of an extraction-condensing heat and power station P_n and the power input to the electrical power system P_e is determined as in a back-pressure heat and power plant, from an equation analogical to (1.12)

$$P_e = P_n \, \eta_{tr} = P_T \, \eta_\varepsilon \, \eta_{tr} = P_T \, (1 - \varepsilon) \, \eta_{tr} \qquad (1.35)$$

The electrical energy generated in this system contains two components: back-pressure energy E_b and condensing energy E_c, but the annual utilisation times for both components are most frequently different. These values are related to the equations (1.14) for back-pressure and (1.36) for condensing power output

$$E_c = P_{cp} \, T_{cp} = P_{cp} \, m_c \, T \qquad (1.36)$$

$$E = E_b + E_c \qquad (1.37)$$

here:
E_b – back-pressure electrical energy, MWh/a;
E_c – condensing electrical energy, MWh/a;
E – total gross electrical energy;
P_{cp} – peak-load electrical power generated in a condensing cycle, MW;
T_{cp} – annual utilisation time of peak-load power generated in a condensing cycle, h/a;
m_c – load factor in the condensing cycle.

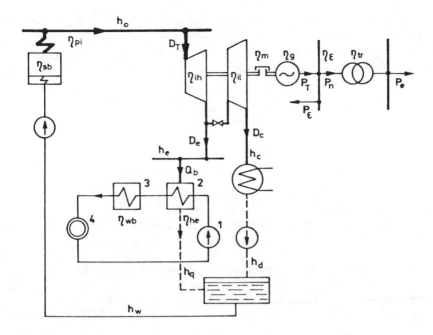

Fig. 1.8   *Heat diagram of an extraction-condensing heat and power plant with a peak-load hot water boiler. 1 – district heating network water pump, 2 – steam-water base-load heat exchanger, 3 – peak-load hot water boiler, 4 – heat receiver*

Whereas the utilisation time of back-pressure output T_{bp} is imposed in view of the utilisation time of extraction heat output, T_{cp} is not related to the heat load and results from the dispatching of condensing power in the power system.

The gross heat output of an extraction-condensing heat and power plant Q_g may be – as in a back-pressure heat and power plant – the sum of heat output from extraction point Q_b and of reduction-cooling valve Q_r (Fig. 1.7). The net heat output is diminished by losses in the heat exchanger, therefore formula (1.13) is applied as in the case of the back-pressure heat and power plant. If, however, instead of a reduction-cooling valve in a heat and power plant delivering heat output in the form of hot water, a peak-load water boiler is introduced (3 in Fig. 1.8), then the net heat output from the heat and power plant amounts to

$$Q_n = Q_b \, \eta_{he} + Q_{wb} \qquad (1.38)$$

here Q_{wb} – heat output of peak-load hot water boiler.

Heat energy delivered from extraction point H_b and from reduction-cooling station H_r, is determined – as in the case of a back-pressure heat and power plant – on the basis of relationships (1.15) and (1.17). If, instead, a peak-load hot water boiler is applied, the net heat energy delivered by the heat and power plant is

$$H_n = H_b \, \eta_{he} + H_{wb} \tag{1.39}$$

here

$$H_{wb} = Q_{wbp} \, T_{wp} \tag{1.40}$$

where:

H_{wb} – heat energy supplied by the peak-load hot water boiler;
Q_{wbp} – peak-load heat output of a hot water boiler;
T_{wp} – annual utilisation time of the peak-load output of a hot water boiler.

Example 1.3

An extraction-condensing turbogenerator set delivers a gross electrical power of $P_T = 25$ MW and a gross heat output $Q_b = 60$ MJ/s in extraction steam with a pressure of $p_e = 0.25$ MPa with inlet parameters of 8·8 MPa, 535 °C, and with condenser pressure of $p_c = 5$ kPa. The efficiencies of the turbogenerator set are: $\eta_{ih} = 0.75$, $\eta_{il} = 0.80$, $\eta_{em} = 0.94$. Calculate the total and specific demands for steam:

(a) at given electrical and heat loads;
(b) for the same electrical power generated in a condensing cycle.

Solution:

With inlet parameters of 8·8 MPa, 535 °C we obtain $h_0 = 3479$ kJ/kg. For an extraction pressure $p_e = 0.25$ MPa, the theoretical (isentropic) enthalpy value is $h_{es} = 2609$ kJ/kg. The theoretical (isentropic) enthalpy drop in the high-pressure part of the turbine amounts to

$$\Delta h_{es} = h_0 - h_{es} = 3479 - 2609 = 870 \text{ kJ/kg}$$

whereas the real enthalpy drop in the high-pressure part is

$$\Delta h_e = \Delta h_{es} \, \eta_{ih} = 870 \cdot 0.75 = 652 \text{ kJ/kg}$$

The enthalpy of extraction steam thus amounts to

$$h_e = h_0 - \Delta h_e = 3479 - 652 = 2827 \text{ kJ/kg}$$

For a condenser pressure $p_c = 5$ kPa, the theoretical enthalpy drop in the low-pressure part of the turbine is $\Delta h_{cs} = 595$ kJ/kg and the real enthalpy drop amounts to

$$\Delta h_c = \Delta h_{cs} \, \eta_{il} = 595 \cdot 0.8 = 476 \text{ kJ/kg}$$

The steam enthalpy before the condenser thus amounts to

$$h_c = h_e - \Delta h_c = 2827 - 476 = 2351 \text{ kJ/kg}$$

From the table of saturated steam at pressure $p_e = 0.25$ MPa we obtain $h_q = 533$ kJ/kg, thus the enthalpy drop in the steam receiver amounts to

$$\Delta h_q = h_e - h_q = 2827 - 533 = 2294 \text{ kJ/kg}$$

The extraction steam flow corresponding to the given heat output is determined from the formula (1.24)

$$D_e = \frac{Q_b}{\Delta h_q} = \frac{60 \cdot 10^3}{2294} = 26.2 \text{ kg/s} = 94.2 \text{ t/h}$$

and the back-pressure power output at the turbogenerator terminals from the formula (1.22) is

$$P_b = D_e \Delta h_e \, \eta_{em} = 26.2 \cdot 652 \cdot 0.94 \cdot 10^{-3} = 16.0 \text{ MW}$$

In view of this, the condensing output from the formula (1.25) amounts to

$$P_c = P_T - P_b = 25 - 16 = 9 \text{ MW}$$

In case (a) with a back-pressure output $P_b = 16$ MW and a condensing output $P_c = 9$ MW, the total demand for steam, according to the formula (1.33), is

$$D_T = \frac{P_T}{\eta_{em}\Delta h} + D_e y_e = \frac{25 \cdot 10^3}{0.94 \cdot 1128} + 26.2 \cdot 0.42$$
$$= 34.7 \text{ kg/s} = 124.8 \text{ t/h}$$

and the specific demand for steam is

$$d_T = \frac{D_T}{P_T} = \frac{124.8}{25} = 4.99 \text{ kg/kWh}$$

In case (b) with a total power output $P_T = 25$ MW and $D_e = 0$, the total demand for steam amounts to

$$D_T = \frac{P_T}{\eta_{em}\Delta h} = \frac{25 \cdot 10^3}{0.94 \cdot 1128} = 23.6 \text{ kg/s} = 84.9 \text{ t/h}$$

and the specific demand for steam is

$$d_T = \frac{D_T}{P_T} = \frac{84.9}{25} = 3.40 \text{ kg/kWh}$$

Example 1.4
Calculate the gross and net annual electrical and heat energy supplied by the extraction-condensing turbogenerator set from Example 1.3 and by a peak-load hot water boiler, connected in series with a base-load exchanger having an efficiency of $\eta_{he} = 0.98$, supplied from the turbine extraction point.

Peak-load heat output values
  – extraction-condensing turbine
  – hot water boiler
annual peak-load utilisation time
  – back-pressure output
  – condensing output
  – hot water boiler heat output
relative consumption of electrical energy for auxiliary requirements $\varepsilon = 0{\cdot}12$.

$$Q_{bp} = 60 \text{ MJ/s}$$
$$Q_{wbp} = 70 \text{ MJ/s}$$

$$T_{bp} = 3500 \text{ h/a}$$
$$T_{cp} = 3000 \text{ h/a}$$
$$T_{wp} = 750 \text{ h/a}$$

Solution:
Taking the parameters given in Example 1.3, the peak-load values of the gross electrical power on the turbogenerator terminals amount to: $P_{bp} = 16$ MW, $P_{cp} = 9$ MW. The annual gross electrical energy E_g is calculated according to the formula (1.37) as the sum of back-pressure energy E_b and of condensing energy E_c

$$E_b = P_{bp}\,T_{bp} = 16 \cdot 3500 \cdot 10^{-3} = 56 \text{ GWh/a}$$
$$E_c = P_{cp}\,T_{cp} = 9 \cdot 3000 \cdot 10^{-3} = 27 \text{ GWh/a}$$
$$E_g = E_b + E_c = 56 + 27 = 83 \text{ GWh/a}$$

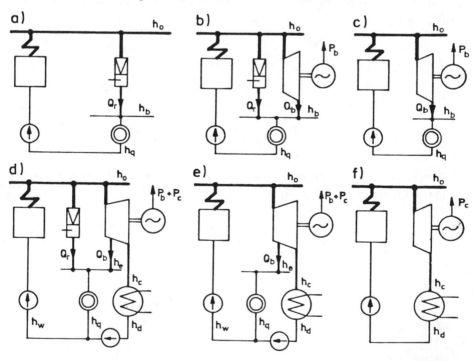

Fig. 1.9 *Simplified diagram to determine the heat-electrical indices:*
(a) low-pressure system (heating plant); (b) back-pressure system with reducer;
(c) back-pressure system; (d) extraction-condensing system with reducer; (e)
extraction-condensing system without reducer; (f) condensing system (power
plant)

In view of this, the annual net electrical energy amounts to

$$E_n = E_g (1 - \varepsilon) = 83 (1 - 0\cdot12) = 73 \text{ GWh/a}$$

The annual gross heat energy H_g is calculated as the sum of heat energy delivered by the turbine H_b and by the hot water boiler H_{wb}

$$H_b = Q_{bp} T_{bp} = 60 \cdot 3500 \cdot 3600 \cdot 10^{-6} = 756 \text{ TJ/a}$$
$$H_{wb} = Q_{wbp} T_{wp} = 70 \cdot 750 \cdot 3600 \cdot 10^{-6} = 189 \text{ TJ/a}$$
$$H_g = H_b + H_{wb} = 756 + 189 = 945 \text{ TJ/a}$$

The annual net heat energy amounts, therefore, to

$$H_n = H_b \, \eta_{he} + H_{wb} = 756 \cdot 0\cdot98 + 189 = 930 \text{ TJ/a}$$

1.3 Heat-electrical indices of heat and power plants

1.3.1 Indices for heat and electrical power and energy

The so called heat-electrical indices result from the relationships between electrical power and energy delivered from a heat and power plant and the corresponding heat output and energy. Lists of these indices in various systems are given in Table 1.1, and the simplified diagrams corresponding to these systems in Fig. 1.9.

These indices can be determined separately for peak-load and separately for annual energy. In the former, the subscript p is added to the main symbol, for example

$$p_p = \frac{P_{bp}}{P_p} , \quad \sigma_p = \frac{P_{bp}}{Q_{bp}} \quad \text{and so on,}$$

and in the latter, the subscript a (Table 1.1).

The following have been accepted for the above indices:

- the index of the share of back-pressure electrical power p_p (p_a for back-pressure electrical energy);
- the index of the share of a back-pressure (extraction) turbine in the peak-load heat output u_{bp} (u_{ba} for heat energy);
- the index of the share of the reducer in the peak-load heat output u_{rp} (u_{ra} for heat energy);
- index of combined production σ (σ_p for peak load, σ_a for annual energy);
- energy index s (s_p for peak-load output, s_a for annual energy).

Relationships occur between the heat-electrical indices

$$u_{ba} + u_{ra} = 1 \tag{1.41}$$

$$\frac{u_{ba} \, \sigma_a \, s_a}{p_a} = 1 \tag{1.42}$$

Table 1.1 List of heat-electrical indices for a heat and power plant

Type of system	Peak-load		Annual energy		Indices for annual energy				
	P_p	Q_p	E	H	$p_a = \dfrac{E_b}{E}$	$u_{ba} = \dfrac{H_b}{H}$	$u_{ra} = \dfrac{H_r}{H}$	$\sigma_a = \dfrac{E_b}{H_b}$	$s_a = \dfrac{H}{E}$
	MW	MJ/s	MWh/a	GJ/a	MWh/MWh	GJ/GJ	GJ/GJ	MWh/GJ	GJ/MWh
Low-pressure only	–	Q_{rp}	–	H_r	–	0	1	–	∞
Back-pressure with reducer	P_{bp}	$Q_{bp} + Q_{rp}$	E_b	$H_b + H_r$	1	<1	<1	σ_a	$> \dfrac{1}{\sigma_a}$
Back-pressure only	P_{bp}	Q_{bp}	E_b	H_b	1	<1	0	σ_a	$\dfrac{1}{\sigma_a}$
Extraction-condensing with reducer	$P_{bp} + P_{cp}$	$Q_{bp} + Q_{rp}$	$E_b + E_c$	$H_b + H_r$	<1	<1	<1	σ_a	$\neq \dfrac{1}{\sigma_a}$
Extraction-condensing without reducer	$P_{bp} + P_{cp}$	Q_{bp}	$E_b + E_c$	H_b	<1	1	0	σ_a	$< \dfrac{1}{\sigma_a}$
Condensing only	P_{cp}	–	E_c	–	0	–	–	–	0

The share index of the back-pressure (extraction) turbine $u_{bp} = Q_{bp}/Q_p$, referring to the peak-load heat output, is frequently found in literature under the name of the combined base-load factor and defined with the symbol α. This should not, however, be mistaken with the combined production index $\sigma_p = P_{bp}/Q_{bp}$, which depends solely on the steam inlet and outlet parameters and on the efficiencies of the turbogenerator set.

Example 1.5
Calculate the heat-electrical indices for the peak-load power output and annual energy delivered from a heating unit, which consists of an extraction-condensing turbogenerator set with an electrical power of $P_T = 25$ MW as in Examples 1.3 and 1.4, and of a peak-load hot water boiler with a heat output of $Q_{wbp} = 70$ MJ/s as in Example 1.4.

Solution:
A peak-load hot water boiler as in Fig. 1.8 is introduced into the extraction-condensing cycle (Fig. 1.9d) instead of the reducer and the heat-electrical indices are calculated as follows:
  for peak-load power output

$$p_p = \frac{P_{bp}}{P_{bp} + P_{cp}} = \frac{16}{16 + 9} = 0.640$$

$$u_{bp} = \frac{Q_{bp}}{Q_{bp} + Q_{wbp}} = \frac{60}{60 + 70} = 0.462$$
$$u_{rp} = 1 - u_{bp} = 1 - 0.462 = 0.538$$

$$\sigma_p = \frac{P_{bp}}{Q_{bp}} = \frac{16}{60} = 0.267$$

$$s_p = \frac{Q_{bp} + Q_{wbp}}{P_{bp} + P_{cp}} = \frac{130}{25} = 5.20$$

for annual energy

$$p_a = \frac{E_b}{E_b + E_c} = \frac{56}{56 + 27} = 0.675$$

$$u_{ba} = \frac{H_b}{H_b + H_{wb}} = \frac{756}{756 + 189} = 0.800$$
$$u_{ra} = 1 - u_{ba} = 1 - 0.800 = 0.200$$

$$\sigma_a = \frac{E_b}{H_b} = \frac{56\,000}{756} = 74.1 \text{ kWh/GJ}$$

$$s_a = \frac{H_b + H_{wb}}{E_b + E_c} = \frac{945}{83} = 11.4 \text{ GJ/MWh}$$

1.3.2 Efficiency of a combined heat and power plant

The total efficiency of a combined heat and power plant is defined as the relationship of energy delivered (gross or net) in both forms (electrical and heat energy) and that supplied in the fuel. Thus, referring the net electrical power delivered P_n and the net heat output Q_n to the heat output suplied to the boilers Q_0, we obtain the total net efficiency

$$\eta_n^{CH} = \frac{P_n + Q_n}{Q_0} = \frac{P_T \eta_\varepsilon + Q_g \eta_{he}}{Q_0} \qquad (1.43)$$

where:

P_T – electrical power at the turbogenerator terminals (gross);

Q_g – heat output delivered to the consumers (gross);

$\eta_\varepsilon = 1 - \varepsilon$ (as in 1.12);

η_{he} – efficiency of heat exchanger at the turbine outlet.

If the heat supplied in the fuel to the steam boilers in the combined heat and power plant Q_0 is divided into two parts, the first of which Q_{0e} corresponds to the conversion into electrical power and the second Q_{0h} to the conversion into heat output, this division corresponds to the following partial efficiencies:

– partial efficiency of the net generation of electrical power

$$\eta_{ne}^{CH} = \frac{P_n}{Q_{0e}} = \eta_{sb}\eta_{pi}\eta_c\eta_{em}\eta_\varepsilon \qquad (1.44)$$

– partial efficiency of the net generation of heat output

$$\eta_{nh}^{CH} = \frac{Q_n}{Q_{0h}} = \eta_{sb}\eta_{pi}\eta_{he} \qquad (1.45)$$

here:

η_{sb} – efficiency of the steam boilers;

η_{pi} – efficiency of pipelines;

η_c – efficiency of steam cycle and remaining efficiencies explained in the previous equations.

In a closed back-pressure cycle (Fig. 1.3), the efficiency of the cycle amounts to

$$\eta_{bc} = 1 \qquad (1.46)$$

irrespective of the internal efficiency of the back-pressure turbine

$$\eta_i = \frac{\Delta h_b}{\Delta h_{bs}} \qquad (1.47)$$

as the heat delivered in the output steam is utilised in full in the receivers and no other losses occur in the cycle.

In view of this, the partial efficiency of the net generation of electrical power in a back-pressure heat and power plant amounts to

$$\eta_{ne}^{BP} = \eta_{sb}\eta_{pi}\eta_{em}\eta_\varepsilon \qquad (1.48)$$

In an extraction-condensing cycle (Fig. 1.7), the efficiency of the cycle, taking into account the heat delivered which is not treated as a loss, amounts to

$$\eta_{ce} = \frac{\alpha_c(h_0 - h_c) + \alpha_e(h_0 - h_e)}{\alpha_c(h_0 - h_d) + \alpha_e(h_0 - h_e)} \qquad (1.49)$$

here

$$\alpha_c = \frac{D_c}{D_T} \qquad (1.50)$$

$$\alpha_e = \frac{D_e}{D_T} \qquad (1.51)$$

In view of this, the partial efficiency of the net generation of electrical power, in an extraction-condensing heat and power plant amounts to

$$\eta_{ne}^{EP} = \eta_{sb}\eta_{pi}\eta_{ce}\eta_{em}\eta_{\varepsilon} \qquad (1.52)$$

Depending upon the relative shares of extraction steam α_e and condensing steam flows α_c, the following operating conditions of an extraction-condensing turbine are determined:
- condensing only at $\alpha_e = 0$; efficiency of this cycle η'_{ce} equals the efficiency of the condensing cycle

$$\eta'_{ce} = \eta_{cc} = \frac{h_0 - h_c}{h_0 - h_d} \qquad (1.53)$$

- almost solely back-pressure operation at $\alpha_c \approx 0$; the efficiency of this cycle η''_{ce} is almost equal to that of the back-pressure cycle

$$\eta''_{ce} \approx \eta_{bc} = \frac{h_0 - h_e}{h_0 - h_e} = 1 \qquad (1.54)$$

The efficiency of an extraction-condensing cycle is generally contained between the boundary values mentioned

$$\eta_{cc} \leqslant \eta_{ce} \leqslant 1 \qquad (1.55)$$

The influence of the efficiency η_{pi} of the steam pipes connecting the boiler to the turbine is presented in Fig. 1.10. As opposed to the previous figures 1.3 and 1.6, in which it has been assumed that the steam parameters in point 4 correspond to both the outlet from the boiler and inlet to the turbine, two different points have been assumed in the h,s diagram in Fig. 1.10, namely:

- point 4, denoting the steam outlet from the steam boiler superheater, where the pressure is p_{0b} and temperature t_{0b}, which correponds to an enthalpy h_{0b};
- point 4', denoting the inlet of steam to the turbine, where the pressure is p_{0t} and temperature t_{0t}, which corresponds to an enthalpy h_{0t}.

Thus a drop in pressure in the steam pipes $\Delta p_{pi} = p_{0b} - p_{0t}$ and temperature drop in the same pipes $\Delta t_{pi} = t_{0b} - t_{0t}$ cause a drop in steam enthalpy of $\Delta h_0 = h_{0b} - h_{0t}$. Thus the isentropic drop in enthalpy in the back-pressure turbine $\Delta h_{bt} = h_4' - h_7'$ is less than the theoretical drop $\Delta h_{bs} = h_4 - h_7$, hence the efficiency of the steam pipelines is defined with the formula

$$\eta_{pi} = \frac{\Delta h_{bt}}{\Delta h_{bs}} \qquad (1.56)$$

and the internal efficiency of a back-pressure turbine

$$\eta_i = \frac{\Delta h_b}{\Delta h_{bt}} \qquad (1.57)$$

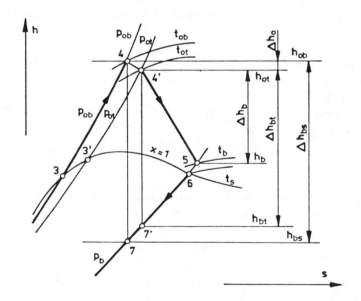

Fig. 1.10   *Influence of pressure and steam temperature drops in the pipelines between boiler and turbine on the enthalpy drops in a back-pressure turbine*

Example 1.6
Calculate the partial efficiency of generating the net electrical and heat energy for an extraction-condensing heat and power plant, fitted with a turbogenerator similar to that in Example 1.3 and a steam boiler with an efficiency of $\eta_{sb} = 0.80$, assuming in addition: pipeline efficiency $\eta_{pi} = 0.99$, consumption of energy for auxiliary demands $\varepsilon = 0.12$ and efficiency of the steam-water heat exchanger $\eta_{he} = 0.98$.

Solution:
Partial efficiency of generating heat energy (net) according to the formula
(1.45)

$$\eta_{nh} = \eta_{sb}\eta_{pi}\eta_{he} = 0.80 \cdot 0.99 \cdot 0.98 = 0.776$$

Efficiency of the extraction-condensing cycle according to the formula
(1.49):
– If

$$D_e = 26.2 \text{ kg/s} = 94.2 \text{ t/h}; \ D_T = 34.7 \text{ kg/s} = 124.8 \text{ t/h}$$

$$\alpha_e = \frac{D_e}{D_T} = \frac{26.2}{34.7} = \frac{94.2}{124.8} = 0.755$$

$$\alpha_c = 1 - \alpha_e = 1 - 0.755 = 0.245$$

then

$$\eta_{ce} = \frac{\alpha_c(h_0 - h_c) + \alpha_e(h_0 - h_e)}{\alpha_c(h_0 - h_d) + \alpha_e(h_0 - h_e)}$$

$$= \frac{0.245 \cdot 1128 + 0.755 \cdot 652}{0.245 \cdot 3343 + 0.755 \cdot 652} = 0.586$$

– If $D_T = D_c = 23.6 \text{ kg/s} = 84.9 \text{ t/h}$

$$\alpha_e = 0; \quad \alpha_c = 1$$

then

$$\eta'_{ce} = \frac{h_0 - h_c}{h_0 - h_d} = \frac{1128}{3343} = 0.337$$

Partial efficiency of generating electrical energy (net) according to the
formula (1.14):
– operating in an extraction cycle

$$\eta_{ne} = \eta_{sb}\eta_{pi}\eta_{ce}\eta_{em}\eta_\varepsilon = 0.80 \cdot 0.99 \cdot 0.586 \cdot 0.94 \cdot 0.88 = 0.384$$

– operating in a condensing cycle

$$\eta'_{ne} = \eta_{sb}\eta_{pi}\eta'_{ce}\eta_{em}\eta_\varepsilon = 0.80 \cdot 0.99 \cdot 0.337 \cdot 0.94 \cdot 0.88 = 0.221$$

Chapter 2

Characteristics of power installations in combined heat and power plants

2.1 Characteristics of back-pressure turbogenerator sets

2.1.1 Parameters and efficiencies of back-pressure turbogenerator sets

A back-pressure turbogenerator set consists of a back-pressure steam turbine and a synchronous generator. In sets with normal speeds, the turbine is connected directly to the generator and has the same rotational speed, which is usually 3000 rpm and in exceptional cases 1500 rpm. In high-speed sets there is a gear between the turbine and generator, which serves to reduce the speed from 7500–12500 rpm on the turbine shaft to 3000 rpm or 1500 rpm on the generator shaft.

The most important parameters of a back-pressure turbine include:

– steam inlet conditions: pressure and temperature of live steam;
– steam outlet conditions: pressure and temperature of back-pressure steam;
– the demand for steam and the capacity of back-pressure outlet;
– mechanical power output at the turbine shaft;
– rotational speed of the turbine.

On the other hand, when the generator is connected to the turbine, its parameters are:

– rated voltage at the terminals,
– electrical power output at the terminals,
– rotational speed of the generator.

The inlet parameters of the turbine are closely related to the outlet parameters of the boilers and together with them are subject to normalisation at specific levels. The steam pressure and temperature levels most frequently noted in back-pressure heat and power plants are listed in Table 2.1, where for each inlet pressure level p_0 two alternative and different inlet temperature values t_0 are given.

Table 2.1 *Steam inlet parameters most frequently noted in back-pressure heat and power plants*

Steam boilers			Steam turbines	
Pressure		Temperature	Pressure	Temperature
Rated	At superheater outlet		At turbine inlet	
MPa	MPa	°C	MPa	°C
1·8	1·6	300	1·5	280
1·8	1·6	350	1·5	330
2·6	2·4	400	2·3	385
2·6	2·4	425	2·3	410
4·2	3·7	425	3·4	410
4·2	3·7	450	3·4	435
7·9	6·9	480	6·4	465
7·9	6·9	500	6·4	485
11·0	9·7	510	8·8	500
11·0	9·7	540	8·8	535
16·0	13·6	540	12·5	535
16·0	13·6	570	12·5	565

The outlet pressure (back-pressure) p_b depends on the steam consumer demand. The values of steam pressure most frequently noted at turbine outlets in back-pressure heat and power plants are listed in Table 2.2. Also given for each level of back-pressure, is its range of variability, which corresponds to the extent to which regulation of the back-pressure at the turbine outlet is possible.

On the other hand, the outlet temperature t_b depends not only on the pressure p_b, but also on the inlet parameters p_0, t_0 and on the internal efficiency of the turbine η_i. In order to illustrate this dependence, a series of steam cycle calculations have been carried out, the results of these being presented in Fig. 2.1. For this, the following data have been assumed:

  – four inlet pressure values: $p_0 = 3\cdot4$; $6\cdot4$; $8\cdot8$; $12\cdot5$ MPa;
  – two temperature values t_0 for each pressure value p_0 as in Fig. 2.1;
  – three back-pressure values: $p_b = 0\cdot25$; $0\cdot5$ and $0\cdot8$ MPa;
  – variability range of internal efficiency $\eta_i = 0\cdot5$–$1\cdot0$.

Table 2.2   *Steam outlet pressures most frequently noted in back-pressure heat and power plants*

Pressure level	MPa	0·12	0·3	0·5	0·7	1·0
Pressure variability range	MPa	0·07–0·25	0·2–0·4	0·4–0·7	0·5–0·9	0·8–1·3
Pressure level	MPa	1·3	1·8	2·4	3·0	4·0
Pressure variability range	MPa	1·0–1·6	1·5–2·1	2·2–2·8	2·8–3·5	3·5–4·3

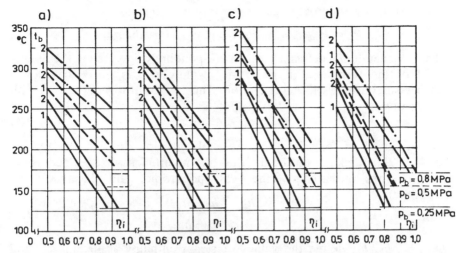

Fig. 2.1   *Temperatures of outlet steam in a heat and power plant depending upon the inlet parameters and efficiency of back-pressure turbine:*

(a)	(b)	(c)	(d)
$p_0 = 3\cdot4\,MPa$	$p_0 = 6\cdot4\,MPa$	$p_0 = 8\cdot8\,MPa$	$p_0 = 12\cdot5\,MPa$
$t_{01} = 410\ ^{\circ}C$	$t_{01} = 465\ ^{\circ}C$	$t_{01} = 500\ ^{\circ}C$	$t_{01} = 535\ ^{\circ}C$
$t_{02} = 435\ ^{\circ}C$	$t_{02} = 485\ ^{\circ}C$	$t_{02} = 535\ ^{\circ}C$	$t_{02} = 565\ ^{\circ}C$

As shown in the diagrams, the outlet temperature t_b drops distinctly with an increase in efficiency η_i; at the same time, of course, this temperature is higher if the back-pressure p_b is higher. The horizontal lines in Fig. 2.1 denote the boundary values to which the temperature t_b can drop, and which result from the steam saturation temperature at the given back-pressure. As can be seen under the given conditions, this boundary is usually attained only with a very high internal turbine efficiency of more than $0\cdot8$.

In practice, the steam temperature t_b at the back-pressure turbine outlet should be higher than the saturation temperature at a given pressure by at least 20–30 K in order to avoid the condensation of steam in the steam pipelines leading from the turbine to the heat consumers.

Apart from the internal efficiency η_i, the following are characteristics of a back-pressure turbogenerator set: mechanical efficiency η_m, transmission gear efficiency η_t and generator efficiency η_g. In accordance with the formula (1.5), the electromechanical efficiency η_{em} constitutes the product of three efficiencies $\eta_m\,\eta_t\,\eta_g$. The notions of efficiency mentioned correspond to the following definitions of the outputs of back-pressure turbogenerator set without regeneration, the scheme of which is presented in Fig. 1.4:

– internal power output of the turbine

$$P_{ib} = D_b\,\Delta h_b \tag{2.1}$$

– mechanical power output on the turbine shaft

$$P_{mb} = P_{ib}\,\eta_m = D_b\,\Delta h_b\,\eta_m \tag{2.2}$$

– electrical power output at generator terminals (gross)

$$P_b = P_{mb} \, \eta_t \, \eta_g = P_{ib} \, \eta_m \, \eta_t \, \eta_g = D_b \, \Delta h_b \, \eta_{em} \qquad (2.3)$$

The regeneration coefficient φ_r, correspondingly increasing the power output of the turbine at a given steam flow D_b, is introduced for a back-pressure turbine with regeneration. Thus, for a turbogenerator set with regeneration

$$P_b = \varphi_r D_b \Delta h_b \, \eta_{em} \qquad (2.4)$$

here – as given in formula (1.3) the real enthalpy drop in the turbine $\Delta h_b = h_0 - h_b$ depends on the internal turbine efficiency η_i, defined as the ratio of drops in enthalpy – the real (polytropic) and the theoretical (isentropic)

$$\eta_i = \frac{\Delta h_b}{\Delta h_{bs}} = \frac{h_0 - h_b}{h_0 - h_{bs}} \qquad (2.5)$$

where h_0, h_b, h_{bs} define the values of steam enthalpy as in Fig. 1.3 and explanations to formulae (1.1–1.5).

2.1.2 Analytical characteristics of back-pressure turbogenerator sets

We call the characteristics of a turbogenerator set the relationship between the heat flow supplied to the turbine set or its increment and the power output delivered by the turbogenerator, which constitutes an independent variable. In view of this, two kinds of characteristics [34] are determined:

(a) energy characteristics of the type

$$Q_T = f(P_T)$$

(b) relative increment characteristics of the type

$$\frac{dQ_T}{dP_T} = f'(P_T)$$

here:

Q_T – heat flow supplied to the turbine set (heat demand of the turbine);
P_T – electrical power output delivered by the turbogenerator.

Energy characteristics serve to calculate the fuel consumption and variable costs in a heat and power station; also to determine the sequence of switching the particular sets on or off when their combined load increases or decreases. The characteristics of relative increments are required to divide the load between co-operating sets.

The energy characteristic of a back-pressure turbogenerator set constitutes the relationship between the heat demand of the turbine Q_{Tb} and two related values, namely electrical power delivered at the generator terminals P_b and heat output supplied from the back-pressure outlet Q_b. This characteristic can be presented in the form of a linear function

$$Q_{Tb} = Q_{ie} + q'_b \, P_b + Q_b \qquad (2.6)$$

where:

Q_{ie} – the electromechanical component of the heat demand of an idle-running turbine set (idle-running losses);

q'_b – specific heat demand for back-pressure generation of electrical power, diminished by the electromechanical losses of the idle-running turbine set.

The value of q'_b can be found from the equation for the balance of power of the turbogenerator set with a rated load of P_{br} at the generator terminals

$$Q_{ie} + q'_b P_{br} = P_{br} \frac{1}{\eta_{em}} \tag{2.7}$$

where $\eta_{em} = \eta_m \, \eta_t \, \eta_g$ – electromechanical efficiency of the turbogenerator set. Hence

$$q'_b = \frac{1}{\eta_{em}} - \frac{Q_{ie}}{P_{br}} = \frac{1}{\eta_{em}} - q_{ie} \tag{2.8}$$

$$q_{ie} = \frac{Q_{ie}}{P_{br}} \tag{2.9}$$

where q_{ie} – the electromechanical component of the specific heat demand of an idle-running turbine set.

A relationship called the internal characteristic exists between the electrical power output P_b of a back-pressure turbogenerator set and the heat output Q_b supplied by this set

$$Q_b = Q_{ih} + \frac{1 - x_{ih}}{\sigma_r} P_b \tag{2.10}$$

here

$$x_{ih} = \frac{Q_{ih}}{Q_{br}} \tag{2.11}$$

$$\sigma_r = \frac{P_{br}}{Q_{br}} \tag{2.12}$$

where:

Q_{ih} – the heat component of the demand for heat of an idle-running turbine set (idle-running losses);

x_{ih} – the relative value of the heat component of losses when running idle;

σ_r – rated value of combined index;

Q_{br} – rated heat output delivered at the back-pressure outlet.

The true energy characteristics of turbogenerator sets are non-linear in view of the dependence of their efficiency on load, however, in approximation, a linear flow of characteristics, both external and internal, can be accepted.

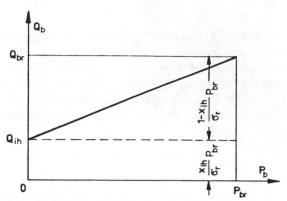

Fig. 2.2   *Internal characteristics of a back-pressure turbogenerator set*

Fig. 2.2 presents the internal characteristic of a back-pressure turbogenerator set in the form of a linear function, defined by the formula (2.10). Substituting the internal characteristic (2.10) into the energy (external) characteristic (2.6), we obtain

$$Q_{Tb} = Q_{ie} + q'_b P_b + Q_{ih} + \frac{1 - x_{ih}}{\sigma_r} P_b \qquad (2.13)$$

and after rearranging

$$Q_{Tb} = Q_i + c_{Tb} P_b \qquad (2.14)$$

where

$$Q_i = Q_{ie} + Q_{ih} \qquad (2.15)$$

$$c_{Tb} = \frac{1}{\eta_{em}} + \frac{1}{\sigma_r} - q_i \qquad (2.16)$$

$$q_i = q_{ie} + q_{ih} = q_{ie} + \frac{x_{ih}}{\sigma_r} = \frac{Q_i}{P_{br}} \qquad (2.17)$$

where:
Q_i – total heat demand of an idle-running turbine set (losses for idle-running);
c_{Tb} – total relative increment of heat demand;
q_i – specific heat demand of an idle-running turbine set;
q_{ih} – heat component of the specific heat demand of an idle-running turbine set.

Fig. 2.3 presents the energy (external) characteristic of a back-pressure turbogenerator set in the form of a linear function, defined by the formula (2.14). The total heat demand Q_{Tb} of the turbine set resulting from this characteristic includes both the heat flow necessary to generate back-pressure electrical power output P_b and the heat output delivered from the back-

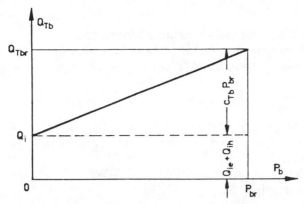

Fig. 2.3   *External characteristics of a back-pressure turbogenerator set*

pressure outlet Q_b. The heat flow Q_{Tb} supplied to the turbine set can therefore be divided into two components Q_{Te} and Q_{Th}, corresponding to the electrical and heat outputs of the turbogenerator set, where

$$Q_{Tb} = Q_{Te} + Q_{Th} \tag{2.18}$$

$$Q_{Te} = Q_{ie} + q'_b P_b \tag{2.19}$$

$$Q_{Th} = Q_{ih} + \frac{1 - x_{ih}}{\sigma_r} P_b = Q_b \tag{2.20}$$

The characteristic of relative increments in heat demand by a back-pressure turbogenerator set is determined by differentiating the energy characteristic (2.14) in relation to P_b, from which we obtain

$$\frac{dQ_{Tb}}{dP_b} = c_{Tb} = q'_b + \frac{1 - x_{ih}}{\sigma_r} \tag{2.21}$$

The relative increment c_{Tb} can also be divided into two component parts c_{Te} and c_{Th}, corresponding to the power and heat output of the turbogenerator set. Here

$$c_{Te} = \frac{1}{\eta_{em}} - q_{ie} = q'_b \tag{2.22}$$

$$c_{Th} = \frac{1}{\sigma_r} - q_{ih} = \frac{1 - x_{ih}}{\sigma_r} \tag{2.23}$$

Apart from energy characteristics of the heat-power type $Q_T = f(P_T)$, characteristics of the demand for steam by the turbine set also apply, i.e. characteristics of the steam-power type. In the case of a back-pressure turbogenerator set, the linear characteristic of steam demand corresponding to the linear energy characteristic (2.14) is in the form

$$D_b = D_{ib} + c_{Db} P_b \tag{2.24}$$

where:

D_{ib} – steam demand of an idle-running turbine set;
c_{Db} – relative increment in steam demand by a back-pressure turbine set;
remaining values – as in formula (1.1).

Example 2.1
Determine the energy characteristics and the characteristics of relative increments in steam demand for a back-pressure turbogenerator set with the following data:
gross rated electric power output $P_{br} = 6{\cdot}0$ MW;
gross rated heat output $Q_{br} = 42{\cdot}0$ MJ/s;
heat demand when running idle $Q_i = 8{\cdot}67$ MJ/s;
electromechanical component of the idle-running losses $Q_{ie} = 0{\cdot}36$ MJ/s;
electromechanical efficiency $\eta_{em} = 0{\cdot}92$.

The heat output Q_b delivered from back-pressure outlet is determined depending upon the electrical output P_b at the generator terminals on the basis of measurements, which gave the following results:

P_b	1·0	2·0	3·0	4·0	5·0	6·0 MW
Q_b	14·23	19·78	25·33	30·88	36·43	41·98 MJ/s

Solution:
The following factors apply:
for energy characteristics – according to formulae (2.8), (2.15), (2.18), (2.19), (2.20)

$$Q_{ih} = Q_i - Q_{ie}$$
$$q'_b = \frac{1}{\eta_{em}} - \frac{Q_{ie}}{P_{br}}$$
$$Q_{Te} = Q_{ie} + q'_b P_b$$
$$Q_{Th} = Q_b$$
$$Q_{Tb} = Q_{Te} + Q_{Th}$$

for characteristics of relative increments – according to formulae (2.22) and (2.23)

$$c_{Te} = q'_b$$
$$c_{Th} = \frac{Q_{br} - Q_{ih}}{P_{br}}$$
$$c_{Tb} = c_{Te} + c_{Th}$$

The characteristics required are listed in Table 2.3.

Table 2.3   Characteristics of the back-pressure turbogenerator set for Example 2.1

P_b MW	Q_{ie} $\frac{MJ}{s}$	Q_{ih} $\frac{MJ}{s}$	Q_i $\frac{MJ}{s}$	Q_b $\frac{MJ}{s}$	q'_b $\frac{MJ}{MJ}$	Q_{Te} $\frac{MJ}{s}$	Q_{Th} $\frac{MJ}{s}$	Q_{Tb} $\frac{MJ}{s}$	c_{Te} $\frac{MJ}{MJ}$	c_{Th} $\frac{MJ}{MJ}$	c_{Tb} $\frac{MJ}{MJ}$
1·0	0·36	8·32	8·68	14·23	1·027	1·39	14·23	15·62	1·027	5·610	6·637
2·0	0·36	8·32	8·68	19·78	1·027	2·41	19·78	22·19	1·027	5·610	6·637
3·0	0·36	8·32	8·68	25·33	1·027	3·44	25·33	28·77	1·027	5·610	6·637
4·0	0·36	8·32	8·68	30·88	1·027	4·47	30·88	35·35	1·027	5·610	6·637
5·0	0·36	8·32	8·68	36·43	1·027	5·50	36·43	41·93	1·027	5·610	6·637
6·0	0·36	8·32	8·68	41·98	1·027	6·52	41·98	48·50	1·027	5·610	6·637

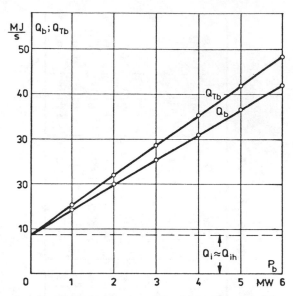

Fig. 2.4   *Characteristics of the turbogenerator set from Example 2.1*

Fig. 2.4 presents the energy characteristics of the turbogenerator set according to the above mentioned relationships – internal $Q_b = f_b(P_b)$ and external $Q_{Tb} = f_T(P_b)$.

2.1.3 Integrated characteristics of back-pressure turbogenerator sets
Apart from the energy characteristics described in the previous section, which, from the point of view of energy economics, are the characteristics of power conversion in turbogenerator sets installed in heat and power plants, also applied are the characteristics of energy conversion, which can be obtained by integrating the energy characteristics from $t = 0$ to $t = T$, with $T = 8760$ h/a.

Thus, integrating equation (2.6) on both sides, we obtain

$$\int_0^T Q_{Tb}\, dt = Q_{ie}\, T_w + q_b' \int_0^T P_b\, dt + \int_0^T Q_b\, dt \qquad (2.25)$$

that is to say

$$H_{Tb} = Q_{ie}\, T_w + q_b'\, E_b + H_b \qquad (2.26)$$

where:
H_{Tb} – heat energy supplied to a back-pressure turbine set during the year (annual heat demand);
T_w – working time of the turbine set during the year, which may be less than $T = 8760$ h/a due to stoppages and repairs;

E_b – electrical energy generated at the generator terminals in a back-pressure cycle;

H_b – heat energy delivered from the back-pressure outlet.

When integrating the right-hand side of the equation (2.6), it is assumed that $q_b' = $ const., that is to say that the electromechanical efficiency of the turboset η_{em} does not depend on the load ($\eta_{em} = $ const.). After further transformations, we obtain the external characteristic of energy conversion in a compound form, analogical to the external characteristic of power conversion defined by the formula (2.14)

$$H_{Tb} = Q_i\, T_w + c_{Tb}\, E_b \tag{2.27}$$

Integrating both sides of the equation (2.10), we obtain the internal characteristic of energy conversion

$$H_b = Q_{ih}\, T_w + \frac{1 - x_{ih}}{\sigma_r}\, E_b \tag{2.28}$$

from which, after dividing the sides by Q_{br} we can obtain the relationship between the utilisation time of the rated heat output and that of the rated electrical output of the turbogenerator set

$$T_{bhr} = x_{ih}\, T_w + (1 - x_{ih})\, T_{ber} \tag{2.29}$$

as well as the reverse relationship

$$T_{ber} = \frac{T_{bhr} - x_{ih}\, T_w}{1 - x_{ih}} \tag{2.30}$$

where:

$T_{bhr} = \dfrac{H_b}{Q_{br}}$ – utilisation time of the rated heat output delivered from the back-pressure outlet;

$T_{ber} = \dfrac{E_b}{P_{br}}$ – utilisation time of the rated electrical power output, generated at the generator terminals.

In order to obtain the analogical relationships between the utilisation time of peak output, electrical P_{bp} and heat Q_{bp} correspondingly, both sides of the equation (2.28) should be divided by Q_{bp}, from which we obtain

$$T_{bhp} = \frac{x_{ih}\, T_w}{n_{qp}} + (1 - x_{ih})\, \frac{n_{pp}}{n_{qp}}\, T_{bep} \tag{2.31}$$

and the reverse relationship

$$T_{bep} = \frac{n_{qp}\, T_{bhp} - x_{ih}\, T_w}{n_{pp}\, (1 - x_{ih})} \tag{2.32}$$

where:

$T_{bhp} = \dfrac{H_b}{Q_{bp}}$ – utilisation time of peak heat output, delivered from the back-pressure outlet;

$T_{bep} = \dfrac{E_b}{P_{bp}}$ – utilisation time of peak electrical power output generated at the generator terminals;

$n_{pp} = \dfrac{P_{bp}}{P_{br}}$ – the ratio of the peak-load electrical output to the rated electrical power output;

$n_{qp} = \dfrac{Q_{bp}}{Q_{br}}$ – ratio of the peak-load heat output to the rated heat output.

It results from the above said that the assumption $T_{bep} = T_{bhp} = T_{bp}$ accepted in formulae (1.14) and (1.15) in section 1.1, is only approximately correct. Both utilisation times would be identical, if there were no heat component of an idle-running turbine set, i.e. if $x_{ih} = 0$. If $x_{ih} \neq 0$, then the utilisation times are different $T_{bep} \neq T_{bhp}$ and the formulae (2.31) and (2.32) should be applied in more accurate calculations.

2.2 Characteristics of extraction-condensing turbogenerator sets

2.2.1 Parameters and efficiencies of extraction-condensing turbogenerator sets

The group of extraction turbogenerator sets includes both back-pressure extraction and extraction-condensing units, with one or two regulated extraction points. Further we have kept to more detailed discussion of the characteristics of extraction-condensing sets with one extraction point and with no transmission gear between the turbine and the generator.

An extraction-condensing turbine may have one or more casings, but in each case, one can distinguish the high-pressure part (HP) from the steam inlet to the regulated extraction point and the low-pressure part (LP) from the extraction point to the steam outlet to the condenser. The most important characteristic parameters of an extraction-condensing turbine include:

– steam inlet conditions: pressure and temperature of live steam;
– steam extraction conditions: pressure and temperature of extraction steam;
– steam outlet conditions: pressure and dryness of steam at the outlet to the condenser;
– maximum output of the regulated extraction point;
– minimum and maximum steam flows in the low-pressure part;
– maximum steam flow in the high-pressure part;
– rotational speed of the turbine.

On the other hand, as regards the synchronous generator connected to the turbine – similar to that mentioned in section 2.1.1 – these are:

– rated voltage at the terminals;
– electrical power output at the terminals;
– rotational speed of the generator.

The inlet parameters p_0, t_0 of extraction turbines are – as in the case of back-pressure turbines – strictly related to the outlet parameters of the boilers and together with them are subject to normalisation at specific levels, which are listed in Table 2.1. The extraction steam pressure p_e depends on the demands of extraction steam consumers. In the case of the supply of extraction steam for process purposes, we find similar pressure values as given in section 2.1.1 for outlet steam from back-pressure turbines. On the other hand, in the case of the supply of extraction steam for heating purposes, i.e. most frequently to supply steam-water exchangers, lower pressure is applied – as a rule less than 0·2 MPa.

As mentioned in section 1.2.2, the internal efficiency of an extraction turbine differs in the high-pressure and in the low-pressure parts. The efficiencies of an extraction-condensing turboset are thus:

– the internal in the high-pressure part η_{ih} and low-pressure part η_{il};
– electromechanical $\eta_{em} = \eta_m \eta_g$.

The above concepts of efficiency correspond to the following definitions of the power output of extraction-condensing turbogenerator set, a diagram of which is given in Fig. 1.7:

– internal power output in the high-pressure part

$$P_{ih} = D_T \Delta h_e \qquad (2.33)$$

– internal power output in the low-pressure part

$$P_{il} = D_c \Delta h_c \qquad (2.34)$$

– internal back-pressure power output, generated on the flow of extraction steam

$$P_{ib} = D_e \Delta h_e \qquad (2.35)$$

– internal condensing power output, generated on the flow of condensing steam

$$P_{ic} = D_c (\Delta h_e + \Delta h_c) = D_c \Delta h \qquad (2.36)$$

– mechanical power output on the turbine shaft

$$P_m = (P_{ih} + P_{il}) \eta_m = (P_{ib} + P_{ic}) \eta_m \qquad (2.37)$$

– electrical power output at the generator terminals

$$P_T = P_m \eta_g = (P_{ih} + P_{il}) \eta_{em} = (P_{ib} + P_{ic}) \eta_{em} \qquad (2.38)$$

where:

D_T – flow of steam supplied to the turbine;
D_e – flow of extraction steam;
D_c – flow of condensing steam;
Δh_e – enthalpy drop in the high-pressure part, as in (1.28);
Δh_c – enthalpy drop in the low-pressure part, as in (1.29).

As results from the formula (2.38), the total electrical power output generated at the generator terminals of the extraction-condensing turbogenerator set can be divided into two component parts in two ways:

(a) depending upon the turbine part – into the output of the high-pressure part P_{HP} and the low-pressure part P_{LP}, thus

$$P_T = P_{HP} + P_{LP} = P_{ih}\,\eta_{em} + P_{il}\,\eta_{em} \qquad (2.39)$$

(b) depending upon the flow of extraction and condensing steam – into the back-pressure output P_b and condensing output P_c, thus

$$P_T = P_b + P_c = P_{ib}\,\eta_{em} + P_{ic}\,\eta_{em} \qquad (2.40)$$

This method of dividing the power output is discussed in greater detail in section 2.2.3, in connection with the energy characteristics of extraction-condensing turbogenerator sets in diagrammatic form.

2.2.2 Analytical characteristics of extraction-condensing turbogenerator sets

The energy characteristics of extraction-condensing turbogenerator sets are slightly more complicated than those of back-pressure units, as in the former there is no functional relationship between the electrical and heat outputs delivered from the extraction point, analogically to equation (2.10). In view of this, the relationship between the heat demand by the turbine Q_T and the electrical power output P_T, as well as the heat output delivered from the extraction point Q_b, cannot be presented in the form of a function with one variable, analogically as in equation (2.14) and remains a function with two variables

$$Q_T = f(P_T, Q_b)$$

This characteristic can thus be written as

$$Q_T = Q_{ic} + q'_b\,P_b + q'_c\,P_c + Q_b \qquad (2.41)$$

where:
Q_{ic} – heat demand when idle-running in a condensing cycle;
P_b – electrical power output generated in a back-pressure cycle;
P_c – electrical power output generated in a condensing cycle;
q'_b – relative increment in specific heat demand for the generation of electrical power in a back-pressure cycle, as in equation (2.6);

q'_c – relative increment in specific heat demand for the generation of electrical power in a condensing cycle;

Q_b – back-pressure heat output delivered from the extraction point.

As the total electrical power output P_T at the generator terminals is the sum of back-pressure output P_b and condensing output P_c, the relationship

$$P_c = P_T - P_b \qquad (2.42)$$

can be inserted in the equation (2.41).

A relationship analogical to equation (2.10) exists, in turn, between the back-pressure electrical power output P_b and back-pressure heat output delivered from the extraction point Q_b, namely

$$Q_b = Q_{ib} + \frac{1}{\sigma_r} \left(1 - \frac{Q_{ib}}{Q_{br}} \right) P_b \qquad (2.43)$$

where:

Q_{ib} – heat component of heat demand in the idle-running of the high-pressure part of the turbine, assumed as the heat demand during idle-running in a back-pressure cycle;

Q_{br} – rated heat output of the turbine, assumed as corresponding to the maximum flow of the extraction steam;

$\sigma_r = \dfrac{P_{br}}{Q_{br}}$ – the rated index of combined production.

Expressing the heat component of losses during idle-running Q_{ib} in the form of a relative value

$$x_{ib} = \frac{Q_{ib}}{Q_{br}} \qquad (2.44)$$

we obtain instead of (2.43) a relationship which corresponds to the internal characteristic of the back-pressure unit

$$Q_b = Q_{ib} + \frac{1 - x_{ib}}{\sigma_r} P_b \qquad (2.45)$$

From this equation we can also define the back-pressure electrical power output P_b depending upon the heat output delivered from the extraction point Q_b

$$P_b = \frac{\sigma_r}{1 - x_{ib}} (Q_b - Q_{ib}) \qquad (2.46)$$

After substituting (2.42) and (2.46) in (2.41), the energy characteristic of the extraction-condensing turbogenerator set is obtained in the form

$$Q_T = Q_{ic} + q_b' \left[\frac{\sigma_r}{1 - x_{ib}} (Q_b - Q_{ib}) \right]$$

$$+ q_c' \left[P_T - \frac{\sigma_r}{1 - x_{ib}} (Q_b - Q_{ib}) \right] + Q_b \qquad (2.47)$$

and after rearranging

$$Q_T = Q_i + q_c' P_T + (1 - \beta) Q_b \qquad (2.48)$$

Here

$$\beta = \frac{\sigma_r(q_c' - q_b')}{1 - x_{ib}} \qquad (2.49)$$

$$Q_i = Q_{ic} + \beta Q_{ib} \qquad (2.50)$$

where:
Q_i – total heat demand during the idle-running of an extraction-condensing turbogenerator set.

If the extraction-condensing turbine operates with the extraction valve closed, as in a solely condensing cycle, and if the extraction pressure regulation system is switched off, then $Q_b = 0$ and $Q_{ib} = 0$ and then we obtain $Q_T = Q_{Tc}$, where

$$Q_{Tc} = Q_{ic} + q_c' P_T \qquad (2.51)$$

This is a form of characteristic analogical to the case of a condensing turbine.

If, on the other hand, the extraction-condensing turbine operates with minimum condensing power output (depending upon the minimum permissible flow of steam through the low-pressure part of the turbine), and in what is for practical purposes a back-pressure cycle, then $P_c \approx 0$ and $P_T \approx P_b$, and then we obtain $Q_T = Q_{Tb}$, where

$$Q_{Tb} = Q_{ic} + q_b' P_b + Q_b \qquad (2.52)$$

or

$$Q_{Tb} = Q_{ic} + q_b' P_b + Q_{ib} + \frac{1 - x_{ib}}{\sigma_r} P_b \qquad (2.53)$$

and after rearranging

$$Q_{Tb} = Q_{ic} + Q_{ib} + c_{Tb} P_b \qquad (2.54)$$

where the relative increment c_{Tb}, related to the generating of back-pressure electrical power output, amounts as in the case of the back-pressure unit, to

$$c_{Tb} = q_b' + \frac{1 - x_{ib}}{\sigma_r} \qquad (2.55)$$

As derived from the formula (2.48), at the given heat output Q_b, delivered from the extraction point, the energy characteristic of the extraction-condensing turbogenerator set is a linear function

$$Q_T = Q_i + c_{Te}\,P_T + (1 - \beta)\,Q_b \qquad (2.56)$$

where $c_{Te} = q'_c$ − the relative increment of heat demand related to the generation of condensing electrical power.

The characteristic of relative increments of heat demand by an extraction-condensing turbogenerator set is determined by the differentiation of the energy characteristic in equation (2.56) with respect to P_T at a constant Q_b. Therefore

$$\frac{dQ_T}{dP_T} = c_{Te} = q'_c \qquad (2.57)$$

Here one immediately obtains the relative increment c_{Te} related to the generation of condensing electrical power.

It should be noted that when dividing the electrical load between power plants working together, the relative increment c_{Te} according to formula (2.57) is taken into account for extraction-condensing turbogenerator sets only in the range of power from the rated output to the minimum condensing one. On the other hand, in the range of power below this minimum, i.e. when operation is on back-pressure only, a relative increment of c_{Tb} according to the formula (2.16) is assumed − as for back-pressure units.

Table 2.4 gives the energy characteristic $Q_T = f(P_T, Q_b)$ and relative increment characteristics c_{Te}, c_{Tb} of an extraction-condensing turbogenerator set of 30 MW with inlet parameters of 8·8 MPa, 535 °C.

Apart from energy characteristics of the heat-power type in respect of extraction-condensing turbogenerator sets, characteristics of the steam-power type obtained from equation (1.33) can also be applied, taking into consideration the losses from idle-running. Such characteristics of the demand for steam by an extraction-condensing turbine set, corresponding to the linear energy characteristic in equation (2.56), has the form

$$D_T = D_i + c_{De}\,P_T + y_e\,D_e \qquad (2.58)$$

where:
D_i − total steam demand by an idle-running turbine set;
c_{De} − relative increment of demand for steam by an extraction-condensing turbine set;
remaining values − as in formula (1.33).

Table 2.4   Characteristics of the extraction-condensing turbogenerator set

$Q_b = 0$

P_T	Q_i	q'_c	Q_T	c_{Te}	c_{Tb}
MW	$\dfrac{\text{MJ}}{\text{s}}$	$\dfrac{\text{MJ}}{\text{MJ}}$	$\dfrac{\text{MJ}}{\text{s}}$	$\dfrac{\text{MJ}}{\text{MJ}}$	$\dfrac{\text{MJ}}{\text{MJ}}$
5·0	4·75	2·676	18·13	2·676	—
10·0	4·75	2·676	31·51	2·676	—
15·0	4·75	2·676	44·89	2·676	—
20·0	4·75	2·676	58·27	2·676	—
25·0	4·75	2·676	71·65	2·676	—
30·0	4·75	2·797	88·66	2·797	—

$Q_b = 0.75 \cdot Q_{b\,max} = 45.3$ MJ/s

P_T	Q_i	q'_c	Q_T	c_{Te}	c_{Tb}
18·5	4·75	2·676	72·40	—	1·009
20·0	4·75	2·676	76·41	2·676	—
25·0	4·75	2·676	89·79	2·676	—

$Q_b = 0.5 \cdot Q_{b\,max} = 30.2$ MJ/s

P_T	Q_i	q'_c	Q_T	c_{Te}	c_{Tb}
MW	$\dfrac{\text{MJ}}{\text{s}}$	$\dfrac{\text{MJ}}{\text{MJ}}$	$\dfrac{\text{MJ}}{\text{s}}$	$\dfrac{\text{MJ}}{\text{MJ}}$	$\dfrac{\text{MJ}}{\text{MJ}}$
—	—	—	—	—	—
11·8	4·75	2·676	48·42	—	1·009
15·0	4·75	2·676	56·99	2·676	—
20·0	4·75	2·676	70·37	2·676	—
25·0	4·75	2·676	83·75	2·676	—
—	—	—	—	—	—

$Q_b = Q_{b\,max} = 60.4$ MJ/s

P_T	Q_i	q'_c	Q_T	c_{Te}	c_{Tb}
—	—	—	—	—	—
24·3	4·75	2·676	93·96	—	1·009
25·0	4·75	2·676	95·83	2·676	—

2.2.3 Graphic characteristics of extraction-condensing turbogenerator sets

The energy characteristics of extraction-condensing turbogenerator sets are presented graphically by means of diagrams:

- of the heat-power type, corresponding to the function $Q_T = f(P_T, Q_b)$ according to the formula (2.48);
- of the steam-power type, corresponding to the function $D_T = f(P_T, D_e)$ according to the formula (2.58).

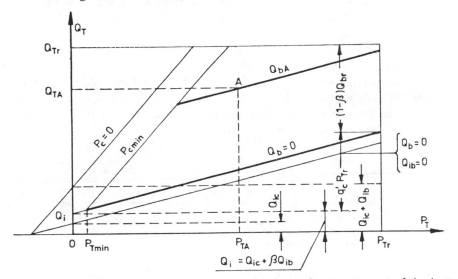

Fig. 2.5 *Characteristics of an extraction-condensing turbogenerator set of the heat-power type*

Fig. 2.5 presents the characteristics of the heat-power type of an extraction-condensing turbogenerator set. In accordance with the formula (2.48), two independent variables occur here: electrical power output P_T at the generator terminals – this being the total of the back-pressure output P_b and condensing output P_c, as in the formula (2.40) – and back-pressure heat output Q_b delivered in the extraction steam.

The electrical power output P_T is determined on the axis of abscissae; heat output Q_b – by means of a series of straight lines parallel to the line $Q_b = 0$, which corresponds to the operating of the turbine set in a solely condensing cycle with a closed extraction valve, but with the extraction pressure regulation system operating. The rated heat flow Q_{Tr} according to the formula (2.48) is expressed by the relationship

$$Q_{Tr} = Q_i + q'_c P_{Tr} + (1 - \beta) Q_{br} \qquad (2.59)$$

Another limitation of the diagram is the line $Q_b = 0$, $Q_{ib} = 0$, which corresponds to the operating of the turbine set in a solely condensing cycle but with the extraction pressure regulation system switched off, when in accordance with the formula (2.51) $Q_T = Q_{Tc}$ and the idle-running heat demand amounts to Q_{ic}, if $P_T = 0$. If, on the other hand, $Q_{ib} \neq 0$, then the idle-running heat demand increases to Q_i, defined by the formula (2.50).

The line $P_{c\ min}$ is the next limitation of the diagram and corresponds to the operation of a turbine set in an almost solely back-pressure cycle, i.e. with minimum condensing power output, resulting from the technical possibilities of diminishing the steam flow D_c in the low-pressure part of the turbine to a minimum value of $D_{c\ min}$. Fig. 2.5 illustrates such a case, in which the minimum power of the turbine set $P_{T\ min} > 0$, as this power is defined by $P_{c\ min}$ at a back-pressure heat output of $Q_b = 0$.

If it were possible to obtain a theoretical line of a solely back-pressure operation $P_c = 0$, then in accordance with the formula (2.54) the heat demand by the turbine set would amount to $Q_T = Q_{Tb}$, and the idle-running heat demand in such a cycle would amount to $Q_{ic} + Q_{ib}$. Further limitations of the diagram are the lines of rated power output P_{Tr} and the rated heat demand by the turbine set Q_{Tr}, defined by the formula (2.59).

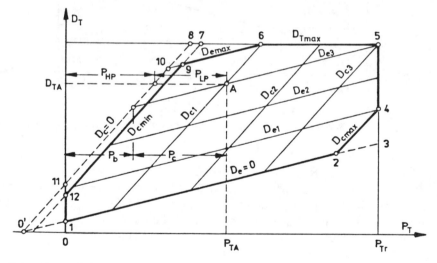

Fig. 2.6   *General characteristics of an extraction-condensing turbogenerator set of the steam-power type*

Fig. 2.6 presents a graphic representation of another kind of extraction-condensing turbogenerator set – the steam-power type. In accordance with the formula (2.58) two independent variables occur in such characteristics: electrical power output P_T at the generator terminals as before, and the flow of extraction steam D_e or the flow of condensing steam D_c.

Electrical power output P_T is defined on the axis of abscissae, and the steam flows D_e and D_c correspondingly by means of straight lines parallel to the lines limiting the characteristics: $D_e = 0$ and $D_c = 0$. Fig. 2.6 presents such a case in which there exist two extreme possibilities of operating a turbine set when idling ($P_T = 0$), namely with a closed extraction valve $D_e = 0$ and $D_c > D_{c\,min}$, and with an open extraction valve $D_e > 0$ and $D_c = D_{c\,min}$.

This is the most general case of characteristics in which the following limitations occur:

1–2   operating line which is solely condensing where $D_e = 0$;
2–4   line of maximum capacity of low-pressure part $D_{c\,max}$;
4–5   line of rated power output P_{Tr};
5–6   line of maximum capacity of high-pressure part $D_{T\,max}$;
6–9   line of maximum extraction capacity $D_{e\,max}$;
9–12 line of minimum capacity of low-pressure part $D_{c\,min}$;
12–1 line of idling $P_T = 0$.

With a power output of $P_T = P_{TA}$ and extraction capacity $D_e = D_{e3}$ the rate of condensing steam flow amounts to $D_c = D_{c1}$, and the demand for steam by the turbine set to $D_T = D_{TA}$. As shown in Fig. 2.6, the total power output of the turbogenerator P_{TA} can be defined as the sum of power generated in both parts of the turbine $P_{HP} + P_{LP}$ in accordance with formula (2.39), or as the

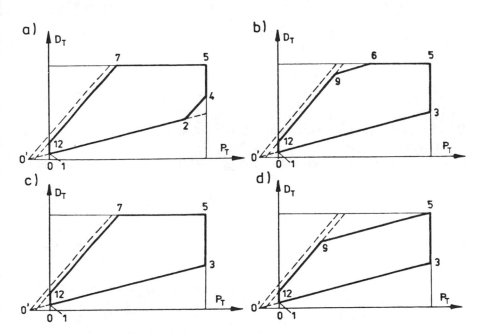

Fig. 2.7   *Specific cases of characteristics of an extraction-condensing turbogenerator set: (a) without limitation $D_{e\,max}$; (b) without limitation $D_{c\,max}$; (c) without limitations $D_{e\,max}$, $D_{c\,max}$; (d) with steam flow $D_{T\,max}$ restricted to point P_{Tr}*

sum of back-pressure and condensing power $P_b + P_c$ in accordance with the formula (2.40).

Fig. 2.7 presents four specific cases of characteristics of the steam-power type, which differ from the general form of this type of characteristics from Fig. 2.6 in the lack of certain limitations, namely:

- where there is no limitation $D_{e\ max}$, instead of three lines: 5–6, 6–9 and 9–12, there are only two: 5–7 and 7–12;
- where there is no limitation $D_{c\ max}$, instead of three lines: 1–2, 2–4 and 4–5 there are only two: 1–3 and 3–5;
- where both limitations $D_{e\ max}$ and $D_{c\ max}$ are lacking, the graph takes the form of a pentagon 1–3–5–7–12;
- where the limitation $D_{e\ max}$ is such that the capacity $D_{T\ max}$ occurs only in p. 5 at rated power output, the graph takes the form of a pentagon 1–3–5–9–12.

2.2.4 Integrated characteristics of extraction-condensing turbogenerator sets

The characteristics of energy conversion of an extraction-condensing turbogenerator set are found by integrating the energy characteristics defined by the formula (2.41)

$$\int_0^T Q_T\,dt = Q_{ic}\,T_w + q_b' \int_0^T P_b\,dt + q_c' \int_0^T P_c\,dt + \int_0^T Q_b\,dt \qquad (2.60)$$

That is to say

$$H_T = Q_{ic}\,T_w + q_b'\,E_b + q_c'\,E_c + H_b \qquad (2.61)$$

where:

H_T – heat energy supplied to an extraction turbine set during the period of one year (annual heat demand);

E_c – electrical energy generated at the generator terminals in a condensing cycle;

H_b – heat energy delivered from the extraction point (back-pressure heat energy).

After further transformations we obtain the characteristics of energy conversion in a three-membered form, analogical to the characteristics of power conversion defined by the formula (2.48)

$$H_T = Q_i\,T_w + c_{Te}\,E_T + (1 - \beta)H_b \qquad (2.62)$$

Here

$$E_T = E_b + E_c \qquad (2.63)$$

where E_T – total electrical energy generated at the generator terminals.

If an extraction-condensing turbine set operates for a period of working time T_{wc} with the extraction valve closed and extraction pressure regulation switched off, i.e. in a solely condensing cycle, then during that time $Q_b = 0$ and $Q_{ib} = 0$ and we obtain

$$H_{Tc} = \int_0^{Twc} Q_{Tc}\, dt = Q_{ic} T_{wc} + c_{Te} E_c \qquad (2.64)$$

If, on the other hand, an extraction-condensing turbine set operates for a period of working time T_{wb} with the condensation switched off almost completely, i.e. in an almost solely back-pressure cycle, then $P_c \approx 0$, $P_T \approx P_b$ and we obtain

$$H_{Tb} = \int_0^{Twb} Q_{Tb}\, dt = (Q_{ic} + Q_{ib}) T_{wb} + c_{Tb} E_b \qquad (2.65)$$

2.3 Characteristics of steam boilers

2.3.1 Parameters and efficiency of steam boilers

Steam boilers applied in combined heat and power plants are as a rule water-tube boilers. Stoker fired and pulverised-fuel boilers can burn solid fuels (coal or lignite). Gas or oil-burning boilers can also be used. The most important parameters of a steam boiler include:

– live steam conditions, i.e. pressure and temperature at the superheater outlet and in the case of a boiler with resuperheater, the pressure and temperature of steam at the resuperheater outlet;
– boiler steaming capacity: maximum constant, economic, and minimum;
– water parameters at the boiler inlet, i.e. pressure and temperature of feed water;
– parameters of air and exhaust gases, i.e. temperature of inlet air and pressure and temperature of exhaust gases leaving the boiler.

The boiler outlet parameters are strictly related to the turbine inlet parameters and together with them are subject to normalisation at specific levels. The boiler steaming capacity is adapted to the demand for steam by the turbine and other consumers of live steam. The temperature of feed water is related to the temperature of feed water heating in regeneration heaters supplied by steam from the regulated and non-regulated bleeding points of the turbine.

The rated hourly fuel consumption of the boiler can be calculated as follows

$$B_{sbr} = \frac{D_{sbr}(h_{0b} - h_{fw})}{H_f \eta_{sb}} \qquad (2.66)$$

here:

D_{sbr} – rated steaming capacity of the steam boiler;
h_{0b} – steam enthalpy at the boiler outlet;
h_{fw} – feed water enthalpy at the boiler inlet;
H_f – lower calorific value of the fuel;
η_{sb} – steam boiler efficiency.

2.3.2 Energy characteristics of steam boilers

As in the case of the turbogenerator set, the characteristics of a steam boiler cover the relationship between the heat flow supplied to the boiler in the fuel or its increment, and the heat output delivered by the boiler which constitutes an independent variable. As previously, two kinds of characteristics are determined:

– energy characteristics of the type

$$Q_{SB} = f(Q_{sb})$$

– characteristics of the relative increment type

$$\frac{d Q_{SB}}{d Q_{sb}} = f'(Q_{sb})$$

in which:
Q_{SB} – heat flow supplied to the boiler in the fuel;
Q_{sb} – heat output delivered by the boiler in the steam;
where, in the case of a boiler without a resuperheater

$$Q_{sb} = D_{sb}(h_{0b} - h_{fw}) \qquad (2.67)$$

and in the case of a boiler with a resuperheater

$$Q_{sb} = D_{sb}(h_{0b} - h_{fw}) + D_{rsh}(h_{r2} - h_{r1}) \qquad (2.68)$$

where:
D_{sb} – rate of steam flow through the boiler together with the first superheater (boiler steaming capacity);
D_{rsh} – rate of steam flow through the resuperheater;
h_{0b} – enthalpy of live steam at the first superheater outlet;
h_{fw} – enthalpy of feed water;
h_{r1} – enthalpy of steam at the resuperheater inlet;
h_{r2} – enthalpy of steam at the resuperheater outlet.

The heat flow Q_{SB} supplied to the boiler in the fuel can be related to the boiler output Q_{sb} and to the losses or efficiency of the boiler

$$Q_{SB} = Q_{sb} + \Delta Q_{sb} \doteq Q_{sb} + Q_{sb}\frac{1 - \eta_{sb}}{\eta_{sb}}$$

$$= Q_{sb}\left(1 + \frac{1 - \eta_{sb}}{\eta_{sb}}\right) = \frac{Q_{sb}}{\eta_{sb}} \qquad (2.69)$$

where:

ΔQ_{sb} – heat losses in the steam boiler;

η_{sb} – steam boiler efficiency.

The characteristics of relative increments of the heat demand of a steam boiler are determined by differentiating the energy characteristics in relation to Q_{sb}

$$c_{SB} = \frac{dQ_{SB}}{dQ_{sb}} = 1 + \frac{d}{dQ_{sb}}(\Delta Q_{sb}) \tag{2.70}$$

where:

c_{SB} – relative increment of heat demand by boiler.

Fig. 2.8 presents the energy characteristics of a steam boiler and the resulting characteristics of relative increments, and Fig. 2.9 the characteristics

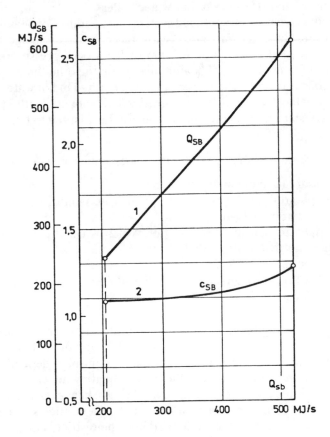

Fig. 2.8   *Characteristics of a steam boiler 1 – energy characteristics, 2 – characteristics of relative increments*

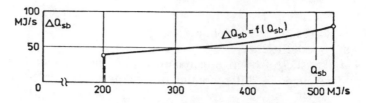

Fig. 2.9   *Characteristics of steam boiler losses*

of boiler losses from which the characteristics of relative increments can also be obtained from formula (2.70).

2.4 Characteristics of hot water boilers

2.4.1 Parameters and efficiency of hot water boilers

Hot water boilers are used in combined systems as the source of peak-load heat output, working together in series with basic heat exchangers of the steam-water type, fed by extraction steam or from back-pressure outlets in heat and power plants. In district heating plants, on the other hand, hot water boilers may constitute the basic source of heat generated in hot water systems. These are usually La Mont type forced-circuit boilers with stoker or pulverised-fuel coal firing. These boilers can also be adapted to burn liquid or gaseous fuel.

The most important parameters of a hot water boiler include:

– water pressure (maximum and working);
– water temperature at the boiler inlet, which may differ when operating normally and at peak load;
– water temperature at the boiler outlet, which is usually the same when operating at normal or peak load;
– rated heat output of boiler;
– rate of water flow through the boiler, which differs under normal and peak load conditions.

The maximum water pressure in a hot water boiler should be adapted to the working pressure in the hot water network fed by the boiler. Under normal working conditions, the temperature at the boiler inlet depends on the temperature of water returning from the district heating network, and under peak-load conditions, on the temperature at the basic heat exchanger outlet in the heat and power plant. The temperature at the boiler outlet is related to the calculated temperature of water discharged into the district heating network.

The rate of water flow through the boiler G_w depends on the heat output Q_{wb} of the hot water boiler, and amounts to

$$G_w = \frac{Q_{wb}}{C_w(\tau_{w2} - \tau_{w1})} \qquad (2.71)$$

here:

C_w – specific heat of water;

τ_{w1} – temperature of water at the boiler inlet;

τ_{w2} – temperature of water at the boiler outlet.

    The rated hourly fuel consumption of the boiler can be calculated as follows

$$B_{wbr} = \frac{Q_{wbr}}{H_f \eta_{wb}} \qquad (2.72)$$

where:

Q_{wbr} – rated heat output of the boiler;

H_f – lower calorific value of the fuel;

η_{wb} – water boiler efficiency.

2.4.2 Energy characteristics of hot water boilers

The characteristics of a hot water boiler constitute – as in the case of a steam boiler – the relationship between the heat flow supplied to the boiler in the fuel or its increment, and the heat output delivered by the boiler which constitutes an independent variable. In view of this, as previously, two kinds of characteristics are determined:

– energy characteristics of the type

$$Q_{WB} = f(Q_{wb})$$

– characteristics of the relative increment type

$$\frac{dQ_{WB}}{dQ_{wb}} = f'(Q_{wb})$$

in which:

Q_{WB} – heat flow supplied to the boiler in the fuel;

Q_{wb} – heat output delivered by the boiler in the hot water.

    Heat flow Q_{WB} supplied to the boiler in the fuel can be related to the heat output delivered Q_{wb} and also to the losses or efficiency of the boiler

$$Q_{WB} = Q_{wb} + \Delta Q_{wb} = Q_{wb} + Q_{wb}\frac{1 - \eta_{wb}}{\eta_{wb}}$$

$$= Q_{wb}\left(1 + \frac{1 - \eta_{wb}}{\eta_{wb}}\right) = \frac{Q_{wb}}{\eta_{wb}} \qquad (2.73)$$

where:

ΔQ_{wb} – heat losses in the water boiler;

η_{wb} – water boiler efficiency.

The characteristics of relative increments of heat demand of a hot water boiler are defined by differentiating the energy characteristics in relation to Q_{wb}

$$c_{WB} = \frac{dQ_{WB}}{dQ_{wb}} = 1 + \frac{d}{dQ_{wb}}(\Delta Q_{wb}) \qquad (2.74)$$

where:
c_{WB} – relative increment of heat demand by boiler.

The energy characteristics of hot water boilers have an analogical form to the characteristics of steam boilers presented in Figs. 2.8 and 2.9.

Chapter 3

Choice of power installations for covering the demands for heat and electrical power in combined systems

3.1 Variation of demands for heat and electrical power

3.1.1 Daily load diagrams of heat and electrical power

The variation of demands for heat and electrical power over the period of a day and a year can be presented in the shape of chronological and ordered load diagrams.

The daily chronological diagram of electrical $P = f(t)$ or heat $Q = f(t)$ load variation presents the corresponding demand in a given industrial plant or power system in the chronological order of loads over the period of one day. Fig. 3.1 shows a typical daily electrical load on a working day in winter, divided into horizontal layers and vertical columns, and Fig. 3.2 the corresponding diagram of the daily heat load. Marked on these diagrams are peak-loads: electrical P_p and heat Q_p, mean (average) loads P_{av}, Q_{av} and the lowest (basic) loads P_{min}, Q_{min}.

In a power system, the electrical peak-load occurs most frequently in the afternoon or evening (the evening peak), although it may also occur in the forenoon. In a district heating system, the heat peak-load usually occurs in the morning (the morning peak). The lowest values of heat and power loads are always during the night.

The daily mean load can be defined on the basis of the daily electrical or heat energy, which is proportional to the area contained between the axis of abscissae and the curve $P = f(t)$ or $Q = f(t)$, and can be defined as an integer

$$E_d = \int_0^{T_d} P(t)dt \qquad (3.1)$$

or

$$H_d = \int_0^{T_d} Q(t)dt \qquad (3.2)$$

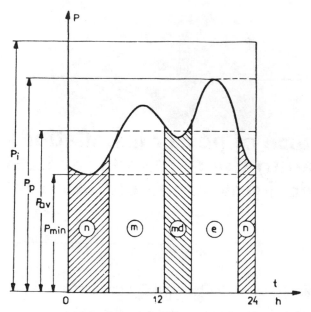

Fig. 3.1   *Daily load diagram of electrical load n – night load, m – morning load, md – mid-day load, e – evening load*

here:
E_d – daily electrical energy, MWh;
H_d – daily heat energy, MJ;
$T_d = 24$ h (twenty-four hours).

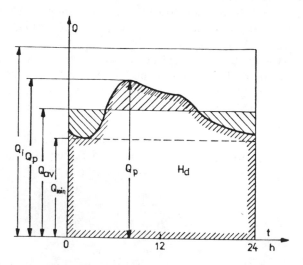

Fig. 3.2   *Daily load diagram of heat load*

In view of this, the corresponding mean load amounts to

$$P_{av} = \frac{E_d}{T_d} = \frac{1}{T_d} \int_0^{T_d} P(t)dt \qquad (3.3)$$

also

$$Q_{av} = \frac{H_d}{T_d} = \frac{1}{T_d} \int_0^{T_d} Q(t)dt \qquad (3.4)$$

The ratio of mean load to peak-load on the daily chronological diagram is called the daily load factor m_d, and the ratio of daily energy to peak-load the daily utilisation time of peak-load output T_{dp}. Relationships occur between these values

– for electrical power output

$$m_{de} = \frac{P_{av}}{P_p} = \frac{E_d}{P_p T_d} = \frac{T_{dpe}}{T_d} \qquad (3.5)$$

– for heat output

$$m_{dh} = \frac{Q_{av}}{Q_p} = \frac{H_d}{Q_p T_d} = \frac{T_{dph}}{T_d} \qquad (3.6)$$

here:
m_{de}, m_{dh} – the daily electrical and heat load factors;
T_{dpe}, T_{dph} – the daily utilisation times of peak-load electrical power and heat output.

The daily ordered diagram of electrical load $P = f(t_d)$ or heat output $Q = f(t_d)$, sometimes called the load duration diagram, presents the daily load, in order of value from the greatest to the smallest, considering the time each lasts. The construction of the daily ordered diagram is given in Fig. 3.3 in respect of the electrical load. The exemplary point B in the ordered diagram is defined by load P and duration time $t_d = t_{d1} + t_{d2}$. Duration time of load P is thus called time t_d, in which load P or greater occurs. Also denoted in Fig. 3.3 are the daily utilisation times of peak-load output T_{dp} and of installed capacity T_{di}.

As the daily energy in the ordered diagram is proportional to the area contained between the axis of abscissae and the curve $P = f(t_d)$ or $Q = f(t_d)$, the integral formulae (3.1) and (3.2), and the denotation of the mean (average) values P_{av} and Q_{av} resulting from the formulae (3.3) and (3.4), are correspondingly applied to the ordered diagrams. The same is the case in respect of the daily electrical m_{de} and heat m_{dc} load factors, defined by the formulae (3.5) and (3.6).

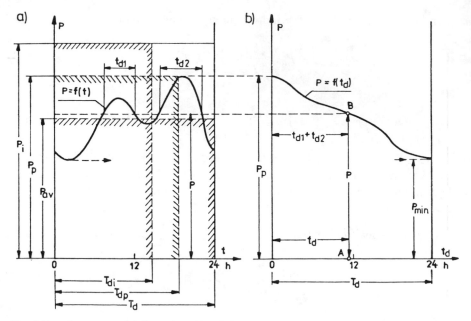

Fig. 3.3   *Structure of daily ordered load diagram of electrical load*
(a) chronological diagram, (b) ordered diagram

3.1.2 Annual load diagrams of heat and electrical power

Setting alongside each other the daily chronological diagrams of consecutive days of the year or typical days which occur a given number of times in the year, annual ordered diagrams of electrical $P = f(t_d)$ and heat $Q = f(t_d)$ loads are denoted, on the basis of which the annual electrical E and heat H energy can be determined

$$E = \int_0^T P(t_d)dt_d \tag{3.7}$$

also

$$H = \int_0^T Q(t_d)dt_d \tag{3.8}$$

where $T = 365 \cdot 24 = 8760$ h – a year.

As previously in the daily diagrams, the annual load factor m_a and the annual utilisation time of peak-load output T_p are defined

– for electrical power output

$$m_{ae} = \frac{E}{P_p T} = \frac{T_{pe}}{T} \tag{3.9}$$

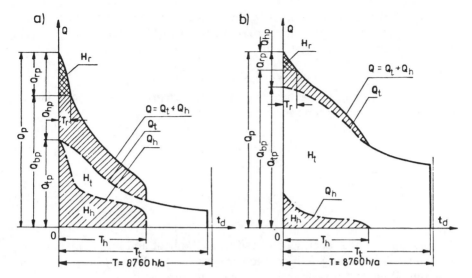

Fig. 3.4 *Annual ordered diagrams of heat load in an industrial heat and power plant:* (a) $u_{tp} = 0.5$; $T_{tp} = 4000$ h/a; (b) $u_{tp} = 0.8$; $T_{tp} = 6000$ h/a

– for heat output

$$m_{ah} = \frac{H}{Q_p T} = \frac{T_{ph}}{T} \qquad (3.10)$$

here:

m_{ae}, m_{ah} – the annual electrical and heat load factors;

T_{pe}, T_{ph} – the annual utilisation times of peak-load electrical power and heat output.

The total heat load Q of a heat and power plant consists of the process (technological) Q_t and heating Q_h loads, which may serve for purposes of space heating alone, as well as for ventilation or air-conditioning purposes. The heat load Q can be covered by steam or hot water. Fig. 3.4 presents two typical ordered diagrams of heat load divided into process (technological) and heating heat. If process and heating peak-loads occur simultaneously, they can be added directly, according to the relationship

$$Q_p = Q_{tp} + Q_{hp} \qquad (3.11)$$

in which:

Q_p – total heat peak-load of the heat and power plant;

Q_{tp} – process heat peak-load;

Q_{hp} – heating peak-load.

Fig. 3.4 also shows the division of heat load Q into the basic part Q_b and peak-load part Q_r. The basic load is covered by the back-pressure or extraction steam from the turbine, and the peak-load from the reducing valve or separate peak-load boilers. The peak-load Q_p thus constitutes the sum of simultaneous peak-loads of the back-pressure (extraction) outlet Q_{bp} and the reducing valve (peak-load source) Q_{rp}. That is to say

$$Q_p = Q_{bp} + Q_{rp} \qquad (3.12)$$

The corresponding annual heat energy values are presented in Fig. 3.4 in the form of areas, which can be summed in two ways

$$H = H_t + H_h \qquad (3.13)$$

or

$$H = H_b + H_r \qquad (3.14)$$

here:
H – total annual heat energy delivered from the heat and power plant;
H_t – annual heat energy delivered for process (technological) purposes;
H_h – annual heat energy delivered for heating purposes;
H_b – annual heat energy delivered from the back-pressure (extraction) outlet;
H_r – annual heat energy delivered from the reducing valve (peak-load source).
The times the particular loads last are:

- for process heat T_t, where $T_t < T$ due to repairs and breakdowns, which may cause a total break in the consumption of heat;
- for heating heat T_h, where in our climatic conditions, the period T_h lasts from October to April and amounts to about 5000 h/a.

The utilisation times of heat peak-loads amount to:

- for process heat

$$T_{tp} = \frac{H_t}{Q_{tp}} \qquad (3.15)$$

- for heating heat

$$T_{hp} = \frac{H_h}{Q_{hp}} \qquad (3.16)$$

- for heat delivered from the back-pressure (extraction) outlet

$$T_{bp} = \frac{H_b}{Q_{bp}} \qquad (3.17)$$

- for heat delivered from the reducing valve (peak-load source)

$$T_{rp} = \frac{H_r}{Q_{rp}} \qquad (3.18)$$

The shares of the particular kinds of heat loads in the total heat peak-load Q_p or in the annual heat demand H, are denoted by the symbol u with the necessary indices. The share thus amounts to:
- for process heat

$$u_{tp} = \frac{Q_{tp}}{Q_p} \tag{3.19}$$

$$u_{ta} = \frac{H_t}{H} \tag{3.20}$$

- for heating heat

$$u_{hp} = \frac{Q_{hp}}{Q_p} \tag{3.21}$$

$$u_{ha} = \frac{H_h}{H} \tag{3.22}$$

- for heat delivered from the back-pressure (extraction) outlet

$$u_{bp} = \frac{Q_{bp}}{Q_p} \tag{3.23}$$

$$u_{ba} = \frac{H_b}{H} \tag{3.24}$$

- for heat delivered from the reducing valve (peak-load source)

$$u_{rp} = \frac{Q_{rp}}{Q_p} \tag{3.25}$$

$$u_{ra} = \frac{H_r}{H} \tag{3.26}$$

In respect of process heat Q_t, usually delivered in the form of steam, more general relationships characterising the variation of heat load in one year cannot be given. This problem should be considered individually for each branch of industry and even for specific larger plants. Investigations have shown only the general dependence of the utilisation time of peak heat demand for process purposes T_{tp} on the number of production shifts. This time amounts to:

in single-shift plants   2000 ÷ 2500 h/a;
in two-shift plants   4000 ÷ 4500 h/a;
in three-shift plants   6000 ÷ 6500 h/a.

Differences in the values given are, however, frequent. Apart from this, it can generally be stated that the diminished utilisation time T_{tp} takes into account the main changes in process load during the day, whereas the distribution of peak-loads on consecutive days of the year is usually fairly even.

As regards heating heat Q_h, usually delivered in the form of hot water, a distinct dependence can be given between the variation of heat load and the time in the heating period, which embraces the winter months from October to April. In the summer months, however, the heat load amounts, on average, to only 5–10% of the peak heat load and is in the main, the heat demand for hot water for utility purposes.

Table 3.1   *Duration times of external temperatures in Central Europe*

External temperature °C	Time h/a	External temperature °C	Time h/a
−27	1	−8	80
−26	1	−7	95
−25	1	−6	112
−24	1	−5	131
−23	2	−4	148
−22	3	−3	171
−21	4	−2	202
−20	5	−1	258
−19	10	0	390
−18	12	1	400
−17	14	2	384
−16	18	3	315
−15	22	4	292
−14	26	5	278
−13	30	6	262
−12	38	7	244
−11	47	8	224
−10	57	9	199
−9	68	10	145
		Total	4690

Table 3.1 gives the length of time the external air temperatures last in the heating season in Central Europe and Fig. 3.5 presents the typical ordered diagram of external temperatures constructed on this basis. The heat demands presented in Fig. 3.4 cover two different, typical cases:

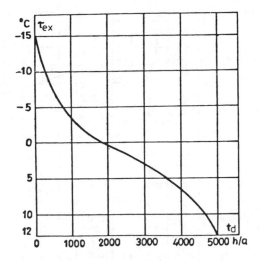

Fig. 3.5   *Annual ordered diagram of external temperatures*

- the utilisation time of the process load is shorter and assumes that average conditions exist in a factory or plant working on two shifts, and that the share of the heating load is relatively high

$$u_{tp} = 0.5 \qquad\qquad u_{hp} = 0.5$$
$$T_{tp} = 4000 \text{ h/a} \qquad T_{hp} = 2000 \text{ h/a}$$

- the utilisation time of the process load is longer and assumes that average conditions exist in a factory or plant working on three shifts, and that the share of the heating load is relatively small

$$u_{tp} = 0.8 \qquad\qquad u_{hp} = 0.2$$
$$T_{tp} = 6000 \text{ h/a} \qquad T_{hp} = 2000 \text{ h/a}$$

The ordered diagram of heating load $Q_h = f(t_d)$ can be drawn up on the basis of an ordered diagram of mean daily external temperatures $\tau_{ex} = f(t_d)$, which is determined for given climatic conditions. A typical ordered diagram of external temperatures is presented in the first quarter of the graph in Fig. 3.6.

The demand for heat most frequently embraces two components, i.e. heat required to cover the heat losses in rooms heated Q_q and heat required in view of the ventilation Q_v. The heat load Q_q changes linearly with the change of external temperature τ_{ex} in the whole range from internal calculated temperature τ_{in} to external rated (calculated) value of τ_{exr}. The heat load Q_v, on the other hand, undergoes linear changes only as regards external calculated temperature for ventilation τ_{exv}, which corresponds to the duration of time $T_v \approx 600$ h/a. Below this temperature limitations of ventilation occur and at $\tau_{exr} < \tau_{ex} < \tau_{exv}$ there is a constant demand $Q_v = \text{const}$. Examples of relationships $Q_q = f_q(\tau_{ex})$ and $Q_v = f_v(\tau_{ex})$ are shown in the second quarter of the graph in Fig. 3.6.

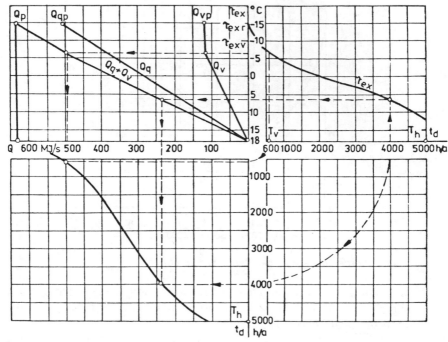

Fig. 3.6   *Structure of annual ordered diagram of heat load*

Example 3.1
Peak heat demand for process purposes from a heat and power plant is $Q_{tp} =$ 120 MJ/s, and peak heating load is $Q_{hp} = 100$ MJ/s. The annual heat demand amounts correspondingly to: process purposes $H_t = 1800$ TJ/a, and heating purposes $H_h = 800$ TJ/a.

Assuming that process and heating peak-loads occur simultaneously, calculate:

(a) the total peak-load of the heat and power plant Q_p, and the total annual demand for heat H;

(b) annual utilisation times of the process load T_{tp}, heating load T_{hp} and total load T_p;

(c) the shares of process and heating heat in the peak-load and annual demand for heat from the heat and power plant.

Solution:
(a)
$$Q_p = Q_{tp} + Q_{hp} = 120 + 100 = 220 \text{ MJ/s}$$
$$H = H_t + H_h = 1800 + 800 = 2600 \text{ TJ/s}$$

(b)
$$T_{tp} = \frac{H_t}{Q_{tp}} = \frac{1800 \cdot 10^6}{120} = 15 \cdot 10^6 \, \text{s/a} = 4160 \, \text{h/a}$$

$$T_{hp} = \frac{H_h}{Q_{hp}} = \frac{800 \cdot 10^6}{100} = 8 \cdot 10^6 \, \text{s/a} = 2220 \, \text{h/a}$$

$$T_p = \frac{H}{Q_p} = \frac{2600 \cdot 10^6}{220} = 11 \cdot 8 \cdot 10^6 \, \text{s/a} = 3280 \, \text{h/a}$$

(c)
$$u_{tp} = \frac{Q_{tp}}{Q_p} = \frac{120}{220} = 0 \cdot 545$$

$$u_{hp} = \frac{Q_{hp}}{Q_p} = \frac{100}{220} = 0 \cdot 455$$

$$u_{ta} = \frac{H_t}{H} = \frac{1800}{2600} = 0 \cdot 692$$

$$u_{ha} = \frac{H_h}{H} = \frac{800}{2600} = 0 \cdot 308$$

Example 3.2
Peak heat load of a heat and power plant is Q_p = 200 MJ/s, of which the process heat share constitutes u_{tp} = 0·75. The annual utilisation time of peak heat load amounts to: process heat T_{tp} = 6000 h/a and heating heat T_{hp} = 2000 h/a.

Calculate:

(a) the peak demand for process heat Q_{tp} and heating heat Q_{hp};
(b) the annual demand for process heat H_t, heating heat H_h, also the total heat H;
(c) resultant utilisation time of peak heat load of heat and power plant T_p;
(d) shares of process and heating heat in total annual heat demand from heat and power plant.

Solution:

(a)
$$Q_{tp} = u_{tp}Q_p = 0 \cdot 75 \cdot 200 = 150 \, \text{MJ/s}$$
$$Q_{hp} = Q_p - Q_{tp} = 200 - 150 = 50 \, \text{MJ/s}$$

(b)
$$H_t = Q_{tp}T_{tp} = 150 \cdot 6000 \cdot 3600 \cdot 10^{-6} = 3240 \, \text{TJ/a}$$
$$H_h = Q_{hp}T_{hp} = 50 \cdot 2000 \cdot 3600 \cdot 10^{-6} = 360 \, \text{TJ/a}$$
$$H = H_t + H_h = 3240 + 360 = 3600 \, \text{TJ/a}$$

(c)
$$T_p = \frac{H}{Q_p} = \frac{3600}{200} \cdot 10^6 = 18 \cdot 10^6 \, \text{s/a} = 5000 \, \text{h/a}$$

(d)

$$u_{ta} = \frac{H_t}{H} = \frac{3240}{3600} = 0\cdot90$$

$$u_{ha} = \frac{H_h}{H} = \frac{360}{3600} = 0\cdot10$$

Example 3.3

Calculate the peak heat demand for heating purposes Q_{hp} and draw an ordered diagram of this demand $Q_h = f(t_d)$ for a heat and power plant supplying an urban-industrial area in which buildings with the following cubature are to be connected up to the heating grid:

Type of building	Cubature (10^6 m³)		
	existing V_1	planned V_2	total V
Dwelling houses (d)	5	10	15
Public utility (p)	3	4	7
Industrial (i)	2	6	8

Specific demand for heat at peak when $\tau_{exr} = -15\ °C$ and $\tau_{in} = +18\ °C$ correspondingly amounts to:

Type of building	Specific heat demand (W/m³)		
	heating q_q	ventilation q_v	total
Dwelling houses (d)	19	1	20
Public utility (p)	20	4	24
Industrial (i)	22	10	32

Connection coefficients depending upon the type of cubature amount to:

Type of building	Coefficient of connected cubature	
	existing c_1	planned c_2
Dwelling houses (d)	0·2	1·0
Public utility (p)	0·8	1·0
Industrial (i)	1·0	1·0

Ordered diagram of external temperatures $\tau_{ex} = f(t_d)$ – according to Fig. 3.5.

Solution:
Peak demand for heating heat
for heating

$$Q_{qp} = V_d \, q_{qd} + V_p \, q_{qp} + V_i \, q_{qi}$$

for ventilation

$$Q_{vp} = V_d \, q_{vd} + V_p \, q_{vp} + V_i \, q_{vi}$$

together

$$Q_{hp} = Q_{qp} + Q_{vp}$$

Cubature connected up

$$V_d = c_{1d}V_{1d} + c_{2d}V_{2d} = (0{\cdot}2 \cdot 5 + 1{\cdot}0 \cdot 10)10^6$$
$$= 11{\cdot}0 \cdot 10^6 \ m^3$$

$$V_p = c_{1p}V_{1p} + c_{2p}V_{2p} = (0{\cdot}8 \cdot 3 + 1{\cdot}0 \cdot 4)10^6$$
$$= 6{\cdot}4 \cdot 10^6 \ m^3$$

$$V_i = c_{1i}V_{1i} + c_{2i}V_{2i} = (1{\cdot}0 \cdot 2 + 1{\cdot}0 \cdot 6) \ 10^6$$
$$= 8{\cdot}0 \cdot 10^6 \ m^3$$

Peak heat demand

$$Q_{qp} = 11 \cdot 19 + 6{\cdot}4 \cdot 20 + 8 \cdot 22 = 513 \ \text{MJ/s}$$
$$Q_{vp} = 11 \cdot 1 + 6{\cdot}4 \cdot 4 + 8 \cdot 10 = 117 \ \text{MJ/s}$$
$$Q_{hp} = 513 + 117 = 630 \ \text{MJ/s}$$

An ordered diagram is drawn by graphic method according to Fig. 3.6. Arrows indicate the sequence of particular points on the corresponding curves.

3.2 Control of heat output delivered in hot water
The heat output delivered from the heat and power plant in the form of hot water to the district heating network depends on the rate of flow of district heating water and the temperature of this water

$$Q = C_w G_w \, (\tau_1 - \tau_2) \tag{3.27}$$

here:
Q – heat output delivered to the district heating network;
C_w – specific heat of water;
G_w – rate of flow of district heating water;
τ_1 – outlet temperature of district heating water;
τ_2 – temperature of return water.

The heat output Q is controlled by adapting the source output to heat demand, which changes depending upon the external temperature. Three types of control are possible:

- qualitative, in which the rate of water flow G_w is constant, whereas the water temperatures τ_1 and τ_2 change;
- quantitative, in which the temperatures τ_1 and τ_2 are constant, whereas the rate of water flow G_w changes;
- qualitative-quantitative, in which the rate of water flow changes in jumps when passing from one temperature variation range to a second one, qualitative regulation taking place within these ranges.

The peak value of heat output Q_p occurs when the temperatures of the district heating water attain calculated peak values τ_{1p} and τ_{2p}

$$Q_p = C_w G_w (\tau_{1p} - \tau_{2p}) \tag{3.28}$$

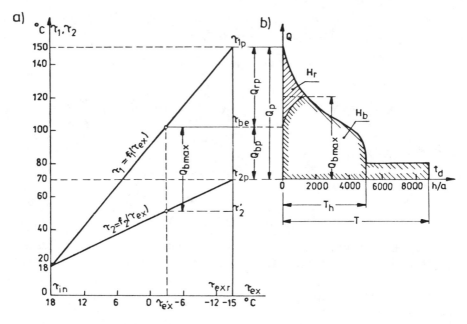

Fig. 3.7   *Qualitative regulation of heat output supplied from a heat and power plant where there is a constant pressure of steam feeding the basic heat exchangers: (a) straight line graph of regulation; (b) ordered diagram of heat output demand*

On the other hand, when $\tau_1 = \tau_2 = \tau_{in}$, then $Q = 0$, thus assuming linear dependences of τ_1 and τ_2 on the external temperature τ_{ex}, a straight line graph of heat output control as in Figs. 3.7 and 3.8 is obtained. At a calculated (rated) external temperature $\tau_{ex} = \tau_{exr}$, the temperatures of the district heating water amount to τ_{1p} and τ_{2p}, and the heat output delivered attains the peak value of Q_p. On the other hand, at a calculated (rated) external temperature

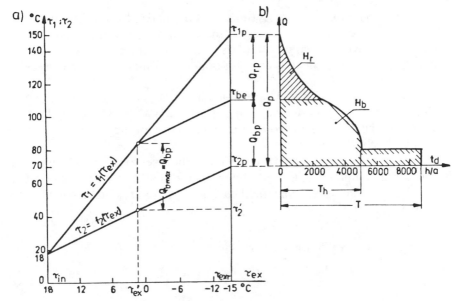

Fig. 3.8 *Qualitative regulation of heat output supplied from a heat and power plant where there is a constant heat output delivered by the basic heat exchangers: (a) straight line graph of regulation; (b) ordered diagram of heat output demand*

$\tau_{ex} = \tau_{in}$ the temperature of the district heating water becomes equalised $\tau_1 = \tau_2$ when the heat output $Q = 0$.

An overall increase in temperature $\tau_1 - \tau_2$ can be carried out in two or more water heaters (heat exchangers), which can be divided into basic heater, supplied with steam from the turbine, and peak-load heaters supplied from reducing-cooling valves. La Mont type water boilers are frequently used instead of peak-load heaters. The temperature of water obtainable behind the basic exchangers τ_{be} depends on the pressure of steam supplying these exchangers from the bleeding point or turbine outlet.

In the case shown in Fig. 3.7, this pressure is constant, thus $\tau_{be} = \text{const}$. In view of this, the division of heat output Q_p, delivered by the heat and power plant in the peak heat period, into output delivered by the basic heaters Q_{bp} and by the peak-load heaters or water boilers Q_{rp}, takes place in the ratio resulting from corresponding temperature differences

$$\frac{Q_{bp}}{Q_{rp}} = \frac{\tau_{be} - \tau_{2p}}{\tau_{1p} - \tau_{be}} \tag{3.29}$$

The greatest heat load of the basic heaters $Q_{b\ max}$, however, occurs at an external temperature of τ_2'. It is easy to demonstrate that $Q_{b\ max} > Q_{bp}$, as $\tau_2' < \tau_{2p}$, whereas

$$\frac{Q_{b\ max}}{Q_{bp}} = \frac{\tau_{be} - \tau_2'}{\tau_{be} - \tau_{2p}} \tag{3.30}$$

The external temperature τ_{exr} is frequently defined as the rated temperature of a severe winter, and that of τ'_{ex} as the rated temperature of an average winter. In an average winter, the basic heaters suffice to cover the heat demand, the peak-load heaters or water boilers becoming essential in a severe winter.

In this case, the division into basic heat energy W_b and peak heat energy W_r is obtained from the ordered diagram of heat demand. A. Olszewski [65] has showed that the dividing line constitutes a reflection of the ordered curve in a correspondingly changed scale.

The ratio of heat output delivered by the turbine set to the total heat output is called the combined base-load factor α in a heat and power plant. Two characteristic values of the combined base-load factor are obtained in the case shown in Fig. 3.7:

– the peak-load value for a severe winter

$$\alpha_p = \frac{Q_{bp}}{Q_p} = \frac{\tau_{be} - \tau_{2p}}{\tau_{1p} - \tau_{2p}} \tag{3.31}$$

– maximum value for an average winter

$$\alpha_{max} = \frac{Q_{b\,max}}{Q_p} = \frac{\tau_{be} - \tau'_2}{\tau_{1p} - \tau_{2p}} \tag{3.32}$$

here $\alpha_{max} > \alpha_p$. The peak-load value of the combined base-load factor α_p corresponds, of course, to the share of the back-pressure heat output u_{bp}, defined in the formula (3.23).

In the case shown in Fig. 3.8, the pressure of steam supplied to the basic heaters changes so that the heat output delivered by these heaters is constant $Q_b = Q_{b\,max} = Q_{bp}$. In an ordered diagram a correspondingly greater area corresponding to peak-load energy W_r is obtained in this case, as the dividing line between W_r and W_b constitutes a horizontal straight line passing through the point with the ordinate Q_{bp}.

A straight lined graph of qualitative control is a simplified one. A more accurate graph of control is obtained by taking into account the non-linear temperature dependencies of τ_1 and τ_2 on the external temperature τ_{ex}. One of the known curvilinear control graphs is that of Dyuskin, which is obtained from the relationship

$$\tau_1 = \tau_{in} + \left[\frac{\tau_{1p} + (2z+1)\,\tau_{2p}}{2(z+1)} - \tau_{in}\right] \left(\frac{\tau_{in} - \tau_{ex}}{\tau_{in} - \tau_{exr}}\right)^{0.8}$$

$$+ \frac{(2z+1)(\tau_{1p} - \tau_{2p})\,G_{wp}}{2(z+1)(\tau_{in} - \tau_{exr})G_w}(\tau_{in} - \tau_{ex}) \tag{3.33}$$

$$\tau_2 = \tau_{in} + \left[\frac{\tau_{1p} + (2z + 1)\,\tau_{2p}}{2(z + 1)} - \tau_{in}\right] \left(\frac{\tau_{in} - \tau_{ex}}{\tau_{in} - \tau_{exr}}\right)^{0\cdot8}$$

$$- \frac{(\tau_{1p} - \tau_{2p})\,G_{wp}}{2(z + 1)(\tau_{in} - \tau_{exr})G_w}\,(\tau_{in} - \tau_{ex}) \tag{3.34}$$

where the coefficient of flow mixing for receiving heat systems is defined by the formula

$$z = \frac{\tau_{1p} - \tau_{qp}}{\tau_{qp} - \tau_{2p}} \tag{3.35}$$

where τ_{qp} – maximum (peak) temperature of water supplied to radiators in inside installations.

Assuming $\tau_{1p} = 150\ ^\circ\mathrm{C}$, $\tau_{qp} = 95\ ^\circ\mathrm{C}$, $\tau_{2p} = 70\ ^\circ\mathrm{C}$, a coefficient of flow mixing $z = 2\cdot2$ is obtained. At a rated temperature of: internal $\tau_{in} = +18\ ^\circ\mathrm{C}$ and external $\tau_{exr} = -15\ ^\circ\mathrm{C}$, the following numerical formulae are obtained from the general formulae (3.33) and (3.34)

$$\tau_1 = 18 + 3\cdot92\,(18 - \tau_{ex})^{0\cdot8} + 2\cdot05\,\frac{G_{wp}}{G_w}\,(18 - \tau_{ex}) \tag{3.36}$$

$$\tau_2 = 18 + 3\cdot92\,(18 - \tau_{ex})^{0\cdot8} - 0\cdot38\,\frac{G_{wp}}{G_w}\,(18 - \tau_{ex}) \tag{3.37}$$

where G_{wp} – peak-load rate of flow of district heating water.

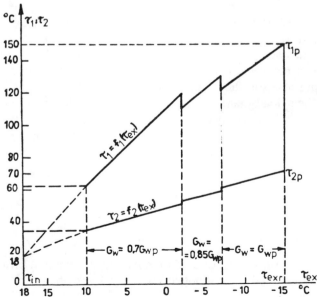

Fig. 3.9   *Curvilinear graph of regulating the qualitative-quantitative heat output delivered from the heat and power plant (after Dyuskin)*

If the control is only qualitative then $G_w = G_{wp}$ = const. On the other hand, qualitative-quantitative control consists in the fact that in some of the external temperature variation ranges, the rated flow of district heating water flow changes so that $G_w < G_{wp}$. Fig. 3.9 presents a curvilinear graph of the control of heat output delivered by the heat and power plant, drawn up on the basis of relationships defined by the formulae (3.36) and (3.37), and here three different values of the rate of flow G_w are assumed

in the first range	$-15 \leqslant \tau_{ex} < -7\,°C$	$G_w = G_{wp}$
in the second range	$-7 \leqslant \tau_{ex} < -2\,°C$	$G_w = 0{\cdot}85\,G_{wp}$
in the third range	$-2 \leqslant \tau_{ex} < +10\,°C$	$G_w = 0{\cdot}7\,G_{wp}$

On the other hand, if the external temperatures are high, in the $+10 \leqslant \tau_{ex} \leqslant +18\,°C$ range, quantitative control is introduced, this consisting in the periodical switching on of the district water pump during the day.

Qualitative-quantitative control enables the cutting of consumption of electrical energy needed to work the district heating water pumps in a heat and power plant. The fault of this method, however, is the variation of the rate of water flow in the district heating network, which causes changes in pressure at the consumer points. Thus the introduction of this method depends on the solution of the problem of automatic control of water pressure in the district heating network nodes.

3.3 Choice of capacity of turbogenerator sets in combined systems

3.3.1 Capacity of back-pressure turbogenerator sets in industrial heat and power plants

The capacity of back-pressure turbogenerator sets, installed in an industrial heat and power plant and delivering heat output in the form of steam mainly for process purposes, is chosen depending upon the heat output demand and the ordered heat load diagram. In accordance with the theoretical bases of combined heat and power systems given in Chapter 1, the gross back-pressure power output at the turbogenerator terminals is calculated on a given turbine steam flow, inlet and outlet steam conditions and the efficiency of turbine set from the formula (1.1)

$$P_b = D_b\,\Delta h_{bs}\,\eta_i\,\eta_{em}$$

here, according to (1.5)

$$\eta_{em} = \eta_m\,\eta_t\,\eta_g$$

The rated steam flow D_{br} of a turbine can be determined on the basis of peak-load heat output Q_{bp}, supplied from the back-pressure outlet, which amounts to

$$Q_{bp} = D_{br}(h_b - h_q) \tag{3.38}$$

here:
h_b – enthalpy of steam at the turbine outlet;
h_q – enthalpy of condensate produced from steam delivered from this outlet.

The peak-load heat output Q_{bp}, which corresponds to the period of time T_r on the ordered diagram (Fig. 3.4), can be determined by applying economic calculations. Below is given the sequence in such calculations, giving the optimum value of the period of time $T_{r\,opt}$, corresponding to minimum annual costs.

In order to determine $T_{r\,opt}$, the energy characteristics of the turbogenerator set should be first defined as in the formula (2.19), in the form

$$Q_{Te} = a\,Q_{br} + b\,P_b \qquad (3.39)$$

here:
Q_{Te} – heat output supplied to the turbine to generate electrical power;
Q_{br} – rated heat output delivered by the turbine;
P_b – electrical power delivered by the generator; a – coefficient of losses when running idle; b – load coefficient.

Thus the annual heat demand of the turbine set amounts to

$$H_{Te} = a\,Q_{br}\,\mu T + b\,E_b \qquad (3.40)$$

where:
H_{Te} – annual heat energy supplied to the turbine for the generation of electrical energy;
μ – relative time the turbine set is operating;
$T = 8760$ h/a – period of one year;
E_b – annual electrical energy supplied by the back-pressure turbine set.

The annual costs of generating energy in a back-pressure cycle amount to

$$K_b = (r + r_o)k_t\,Q_{br} + k_{qT}\,H_{Te} \qquad (3.41)$$

where:
r – annual reproduction rate;
r_o – coefficient of annual fixed operation costs;
k_t – specific investment cost of a back-pressure turbine;
k_{qT} – specific cost of heat delivered to the turbine.

The value of electrical energy generated in a back-pressure cycle

$$K_{Eb} = k_{Eb}\,E_b \qquad (3.42)$$

here k_{Eb} – unit value of back-pressure energy, equal to the unit cost of the supply of electrical energy from a power system loco heat and power plant.

Difference between the value of electrical energy generated and the costs of its generation, i.e. the profits from a combined system amount to

$$-\Delta K = K_b - K_{Eb} = (r + r_o)k_t\,Q_{br} + k_{qT}\,H_{Te} - k_{Eb}\,E_b \qquad (3.43)$$

Assuming next

$$H_{Te} = H_b \, (\delta - 1) \tag{3.44}$$

here δ – coefficient taking into account the increasing demand for heat H_{Te} by the turbine in relation to heat delivered H_b in view of the generation of energy E_b, thus calculating E_b from the equation (3.40)

$$E_b = \frac{H_{Te} - a \, Q_{br} \, \mu T}{b} = H_b \frac{\delta - 1}{b} - \frac{a}{b} Q_{br} \mu T \tag{3.45}$$

and substituting in (3.34), we obtain the equation

$$-\Delta K = (r + r_o) \, k_t \, Q_{br} + k_{qT} \, H_b (\delta - 1) - k_{Eb} \left(\frac{\delta - 1}{b} H_b - \frac{a}{b} Q_{br} \mu T \right) \tag{3.46}$$

Assuming that the peak heat load of the turbine set Q_{bp} is equal to its rated heat output Q_{br}, differentiating the equation (3.46) in relation to Q_{bp}, and equating the first derivative to zero

$$\frac{\partial \Delta K}{\partial Q_{bp}} = 0 \tag{3.47}$$

we obtain

$$(r + r_o) \, k_t + k_{qT} (\delta - 1) \frac{\partial H_b}{\partial Q_{bp}} = k_{Eb} \left(\frac{\delta - 1}{b} \cdot \frac{\partial H_b}{\partial Q_{bp}} - \frac{a}{b} \mu T \right) \tag{3.48}$$

from which after arranging one can calculate

$$\frac{\partial H_b}{\partial Q_{bp}} = \frac{a k_{Eb} \, \mu T + (r + r_o) k_t \, b}{(\delta - 1) \, (k_{Eb} - k_{qT} b)} \tag{3.49}$$

It results from Fig. 3.10 that the duration time of load Q_{bp} amounts to

$$T_r = \frac{\partial H_b}{\partial Q_{bp}} \tag{3.50}$$

Thus for a given ordered diagram $T_{r \, opt}$ and hence Q_{bp} and D_{bp} can be determined from the equations (3.49) and (3.50).

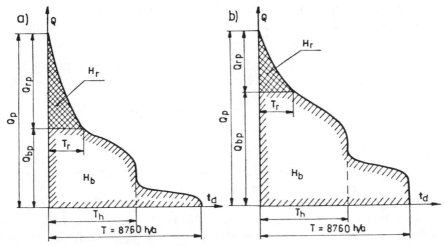

Fig. 3.10   *Annual ordered diagrams of heat and power plant heat load to Example 3.4:*
(a) T_r = 2020 h/a; u_{bp} = 0·45; (b) T_r = 2020 h/a; u_{bp} = 0·65

Example 3.4

Calculate the optimum share $\alpha_p = u_{bp}$ of heat output delivered from the back-pressure outlet in relation to the peak heat load of a heat and power plant which supplies only heating heat, assuming

$a = 0·01$	$\mu T = 5000 \text{ h/a}$
$b = 4·0 \text{ GJ/MWh} = 1·11 \text{ MWh/MWh}$	$k_t = 800 \cdot 10^3 \text{ m.u./MW*}$
$r + r_o = 0·16$	$k_{qT} = 60 \text{ m.u./GJ}$
$\delta = 1·14$	$k_{Eb} = 900 \text{ m.u./MWh}$

Solution:

The duration time $T_{r\,opt}$, which corresponds to the optimum heat output $Q_{b\,opt}$ is calculated from the formula (3.49)

$$T_{r\,opt} = \frac{\partial H_b}{\partial Q_{bp}} = \frac{ak_{Eb}\,\mu T + (r + r_o)\,k_t\,b}{(\delta - 1)\,(k_{Eb} - k_{qT}\,b)}$$

$$= \frac{0·01 \cdot 900 \cdot 5000 \cdot 0·16 \cdot 800 \cdot 10^3 \cdot 1·11}{(1·14 - 1)\,(900 - 60 \cdot 4·0)} = 2020 \text{ h/a}$$

Fig. 3.10 presents two typical annual ordered diagrams of heat load of a heat and power plant, differing in load, that is to say in the annual utilisation time of peak-load

$$T_p = \frac{H}{Q_p} = \frac{H_b + H_r}{Q_{bp} + Q_{rp}}$$

* m.u. = monetary unit

In Fig. 3.10a the utilisation time T_p is less than in Fig. 3.10b. For the calculated value $T_{r\,opt} = 2020$ h/a, if the utilisation time T_p is lower, we obtain a lower value

$$\alpha_{p\,opt} = u_{bp\,opt} = \frac{Q_{bp}}{Q_p} = 0\cdot45$$

and if the utilisation time T_p is higher, we obtain a higher value of $\alpha_{p\,opt} = 0\cdot65$.

Example 3.5
Calculate the total gross electrical power output of a back-pressure turbo-generator set with a rated steam flow D_{br}, so chosen as to correspond to the heat output Q_{bp} delivered from the back-pressure outlet, defined in Example 3.4 for a heat and power plant with a peak-load of $Q_p = 100$ MJ/s. Efficiency of turbine set: $\eta_m = 0\cdot97$; $\eta_t = 1\cdot00$ (without gears); $\eta_g = 0\cdot95$. Enthalpy value of steam and condensate from steam delivered: $h_0 = 3350$ kJ/kg – at the turbine inlet; $h_b = 2720$ kJ/kg – at the turbine outlet; $h_q = 500$ kJ/kg – return condensate.

Solution:
When $u_{bp} = 0\cdot45$ (Fig. 3.10a), the heat output delivered from the back-pressure outlet amounts to

$$Q_{bp} = u_{bp}\,Q_p = 0\cdot45 \cdot 100 = 45 \text{ MJ/s}$$

The rated steam flow of back-pressure turbine set from the formula (3.38)

$$D_{br} = \frac{Q_{bp}}{h_b - h_q} = \frac{45 \cdot 10^3}{2720 - 500} = 20\cdot3 \text{ kg/s} = 73 \text{ t/h}$$

Rated power of back-pressure turbogenerator set accordingly to formula (1.1)

$$P_{br} = D_{br}\,\Delta h_{bs}\,\eta_i\,\eta_{em} = D_{br}(h_0 - h_b)\,\eta_m\,\eta_g$$
$$= 20\cdot3(3350 - 2720)\,0\cdot97 \cdot 0\cdot95 \cdot 10^{-3} \approx 12 \text{ MW}$$

When $u_{bp} = 0\cdot65$ (Fig. 3.10b), the heat output delivered from the back-pressure outlet at the same peak value of heat output Q_p is correspondingly greater and amounts to $Q_{bp} = 65$ MJ/s. The rated steam flow of back-pressure turbine set is then $D_{br} = 29\cdot3$ kg/s = 105 t/h and the rated power of this set $P_{br} \approx 17$ MW.

3.3.2 Type and capacity of turbogenerator sets in district heat and power plants
Back-pressure or extraction-condensing turbogenerator sets, installed in a district heat and power plant, deliver heat output in the form of hot water, mainly for heating purposes.

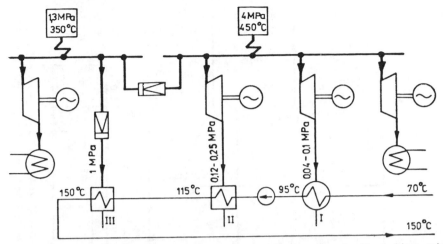

Fig. 3.11 *Heat diagram of a heat and power plant equipped with turbines with diminished vacuum*

Fig. 3.11 presents the heat diagram of a heat and power plant reconstructed from a condensing power plant. Here, the condenser of a turbine operating with a diminished vacuum constitutes the first stage in heating the district heating water. The temperature of this water on entering the heat and power plant at maximum heat load amounts to 70 °C, and after the first heating stage – to 95 °C. The second heating stage is a heat exchanger fed with outlet steam from a back-pressure turbine with a pressure of 120–250 kPa, which can heat the water to about 115 °C. The third stage constitutes a peak-load heat exchanger fed with live steam from the boilers through a reducing-cooling valve, in which the district heating water is heated to the required temperature of 150 °C when leaving the heat and power plant.

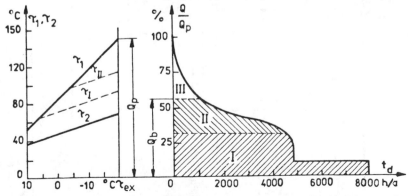

Fig. 3.12 *Three-stage division of heat output and energy generated in a heat and power plant*

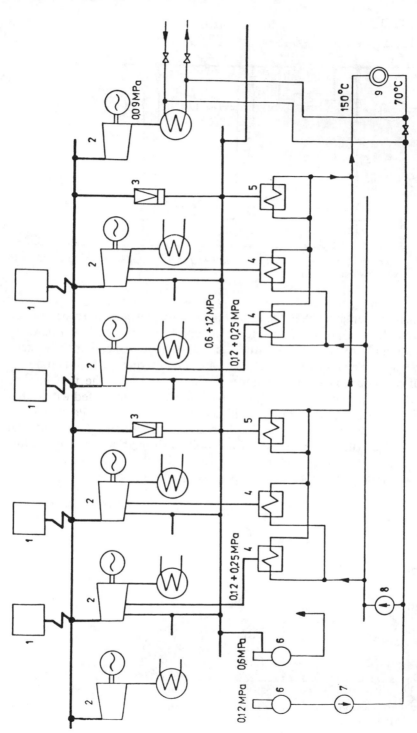

Fig. 3.13   Heat diagram of a heat and power plant with extraction-condensing turbines 1 – high-pressure steam boiler, 2 – steam turbine set, 3 – reducing-cooling valve, 4 – basic heat exchanger, 5 – peak-load heat exchanger, 6 – make-up water tank with deaerator, 7 – make-up water pump, 8 – district heating water pump, 9 – heat receiver

The division of heat load between the stages in the case of three-stage heating of district heating water, is shown in the ordered diagram of heat load of a heat and power plant (Fig. 3.12). The turbine with diminished vacuum operates as the basic one during the whole heating season, and covers the demand for heat in domestic hot water in summer; the back-pressure turbine operates with a shorter utilisation time in the higher layer, and the peak-load heat exchanger operates for only about 1000 h/a when the external temperatures are very low.

The combined base-load factor α, i.e. the ratio of heat output delivered from the turbine, to the maximum heat output delivered from the heat and power plant, amounts to $0.4 - 0.55$, if the demand for heat is solely for heating purposes. Higher values of this coefficient refer to heat and power plants which supply heat not only for the purpose of space heating, but also for process purposes.

Fig. 3.13 presents the heat diagram of a heat and power plant with extraction-condensing turbines. This is a typical collector system in which the total boiler capacity should correspond to the total demand for heat by the turbines and reducing-cooling valves. In the variant with peak-load hot water boilers, the increased utilisation time of high-pressure steam boilers results from the lower output of these boilers.

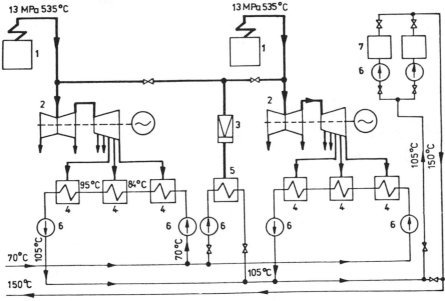

Fig. 3.14   *Heat diagram of a heat and power plant with district heating units containing back-pressure turbine sets and peak-load hot water boilers 1 – high-pressure steam boilers, 2 – district heating back-pressure turbine set, 3 – reducing-cooling valve, 4 – basic heat exchanger, 5 – standby-starting heat exchanger, 6 – district heating water pump, 7 – peak-load hot water boiler*

Faults of the cycle presented in Fig. 3.13 are, among other things, the high parameters of the turbine outlet steam, uneconomic condensing in the turbines over the year, and incorrect utilisation of the high-pressure steam boilers to cover short term peak-loads. The adoption of this system was, however, justified during the period when more efficient condensing power plants with higher inlet parameters were not yet being built.

To increase the profitability of the new, large heat and power plants, modern unit systems with back-pressure turbogenerator sets and peak-load hot water boilers were designed. The choice of back-pressure turbines enables substantial lowering of investment costs and specific fuel consumption. During the summer season and warmer periods of the heating season, these turbines can operate as condensing turbines.

Fig. 3.14 presents a heat diagram of a heat and power plant with back-pressure units. Apart from two heating units, this system contains a reducing-cooling valve common to the two units, which feeds the reserve-starting heat exchanger. In this case, inlet parameters of 12·5 MPa, 535 °C without re-superheating were assumed for district heating turbines which are not destined for condensing work during the heating season.

At present, installations with these parameters have been fully mastered from the point of view of operation and they obtain satisfactory fuel savings with sufficiently high reliability of operation. As the same parameters of live steam are applied at present in newly-built condensing power plants, many elements of the heat diagram which are identical in power plants and in heat and power plants can be utilised. It is planned to increase the inlet parameters in heat and power plants to 16·3 MPa, 550–565 °C.

Chapter 4

Economic effects of combined heat and power generation

4.1 Fuel savings in combined systems

The consumption of fuel for the generation of electrical and heat energy in a combined system comprising a heat and power plant is less than that for the separate generation of electrical and heat energy in independent power and heating plants. The difference in fuel consumption in favour of a combined system, that is to say fuel savings in comparison with separate systems, constitutes the main economic effect of combined production.

To determine the annual fuel saving in the process of the combined generation of electrical and heat energy in a heat and power plant, it is necessary to calculate the annual fuel consumption in a combined system, which may be a back-pressure, extraction-backpressure or extraction-condensing system; also the annual fuel consumption in an equivalent separate system which consists of independent condensing power and heating plants.

4.1.1 Fuel consumption in a back-pressure system

In a back-pressure heat and power plant, the heat diagram of which is presented in Fig. 1.4, the instantaneous fuel consumption (in kg/s or t/h) by the steam boiler B_{sb} is the sum of two components

$$B_{sb} = B_e + B_h \qquad (4.1)$$

where:
B_e – the instantaneous fuel consumption for the generation of electrical power;
B_h – the instantaneous fuel consumption for the generation of heat.

The instantaneous fuel consumption for the generating of the particular kinds of energy can be determined depending upon the heat output delivered from the heat and power plant and the partial efficiency of generating energy

$$B_e = \frac{\sigma_n Q_b}{H_f \eta_{ne}} \qquad (4.2)$$

$$B_h = \frac{Q_n}{H_f \, \eta_{nh}} \qquad (4.3)$$

where:
Q_b – heat output delivered from the back-pressure outlet;
Q_n – net heat output delivered from the heat and power plant;
H_f – the calorific value of the fuel, which for the so-called equivalent coal amounts to 29·3 MJ/kg;
σ_n – net value of combined production index, determined from the formula

$$\sigma_n = \frac{P_{bn}}{Q_b} = \frac{P_b \, \eta_\varepsilon}{Q_b} = \frac{P_b (1 - \varepsilon)}{Q_b} = \sigma_g (1 - \varepsilon) \qquad (4.4)$$

in which:
P_{bn} – net back-pressure electrical power output;
P_b – gross back-pressure electrical power output;
ε – relative power consumption for auxiliary power demand;
σ_g – gross value of combined production index.
   The partial efficiencies of the net energy generation amount to:
for electrical energy

$$\eta_{ne} = \eta_{sb} \cdot \eta_{pi} \cdot \eta_{em} \cdot \eta_\varepsilon \qquad (4.5)$$

for heat energy

$$\eta_{nh} = \eta_{sb} \cdot \eta_{pi} \cdot \eta_{he} \qquad (4.6)$$

where:
η_{sb} – efficiency of steam boiler;
η_{pi} – efficiency of steam pipelines;
$\eta_{em} = \eta_m \cdot \eta_t \cdot \eta_g$ – electromechanical efficiency of the turbogenerator set;
$\eta_\varepsilon = 1 - \varepsilon$ – 'efficiency' of auxiliary power demand;
η_m – mechanical efficiency of the turbine;
η_t – transmission gear efficiency;
η_g – generator efficiency;
η_{he} – efficiency of heat exchanger.
   When calculating the combined production index σ, defining the ratio of electrical power at the generator terminals to the heat output delivered from the back-pressure outlet, the electrical power generated on the steam flow delivered from the turbine bleeding points is also taken into account.
   When calculating the combined net heat output Q_n of the heat and power plant, the heat output Q_r delivered from the reducing-cooling valve is also taken into account, thus

$$Q_n = (Q_b + Q_r) \eta_{he} \qquad (4.7)$$

If, however, a peak-load water boiler serves to cover the peak heat load in the heat and power plant, instead of a reducing-cooling valve, then the combined instantaneous fuel consumption B in the plant, is the sum of three components

$$B = B_{sb} + B_{wb} = B_e + B_h + B_{wb} \tag{4.8}$$

At the same time, only the basic part Q_b of the heat load is assumed to determine the fuel consumption B_h for the generation of heat energy by a steam boiler, whereas the remaining, peak part of the load is taken by a water boiler for which the instantaneous fuel consumption amounts to

$$B_{wb} = \frac{Q_{wb}}{H_f \eta_{wb}} \tag{4.9}$$

where:
Q_{wb} – heat output delivered by the peak-load water boiler;
η_{wb} – efficiency of water boiler.

4.1.2 Fuel consumption in an extraction-condensing system
In an extraction-condensing heat and power plant, the heat diagram of which is presented in Fig. 1.7, the instantaneous fuel consumption by the steam boiler B_{sb} is the sum of three components

$$B_{sb} = B_{eb} + B_{ec} + B_h \tag{4.10}$$

where:
B_{eb} – instantaneous fuel consumption for the generation of back-pressure electrical power;
B_{ec} – instantaneous fuel consumption for the generation of condensing electrical power;
B_h – instantaneous fuel consumption for the generation of heat (from formula 4.3), where

$$B_{eb} = \frac{P_b}{H_f \eta_{eb}} = \frac{\sigma Q_b}{H_f \eta_{eb}} \tag{4.11}$$

$$B_{ec} = \frac{P_c}{H_f \eta_{ec}} \tag{4.12}$$

where:
σ – combined production index as in (4.4);
P_b – back-pressure electrical power generated on steam flow D_e (heat output Q_b delivered from extraction);
P_c – condensing electrical power generated on the steam flow to the condenser D_c.

The partial efficiency of generating electrical energy related to the power at the generator terminals amounts to:
for back-pressure electrical energy

$$\eta_{eb} = \eta_{sb} \cdot \eta_{pi} \cdot \eta_{em} \tag{4.13}$$

for condensing electrical energy

$$\eta_{ec} = \eta_{sb} \cdot \eta_{pi} \cdot \eta_{cc} \cdot \eta_{em} \tag{4.14}$$

where η_{cc} – efficiency of the condensing cycle.

From the formulae (4.11) and (4.12) the specific fuel consumption for the generation of a unit of electrical energy can also be determined correspondingly – back-pressure and condensing

$$b_{eb} = \frac{1}{H_f \cdot \eta_{eb}} \tag{4.15}$$

$$b_{ec} = \frac{1}{H_f \cdot \eta_{ec}} \tag{4.16}$$

here $b_{eb} < b_{ec}$, as $\eta_{eb} > \eta_{ec}$.

As in the case of the back-pressure system, when calculating electrical power P_b and P_c the power generated on the steam flows delivered from the turbine bleeding points is also taken into account, and when calculating the combined net heat output Q_n, the heat output delivered from the reducing-cooling valve is taken into account, in accordance with the formula (4.7).

When using a peak-load water boiler to cover the peak part of the heat load in a heat and power plant (Fig. 1.8), the consumption B_{wb} calculated from the formula (4.9) is added to the consumption B_{sb}, calculated on the basis of the formula (4.10), as in a back-pressure system.

4.1.3 Fuel consumption in a separate system

Equivalent to a heat and power plant, a separate system constitutes: a substituting condensing power plant and a substituting heating plant.

If the heat load of a heat and power plant consists of process heat load Q_t and heating load Q_h, the substituting heating plant would be equipped with low pressure steam boilers with a heat output of not less than Q_t and steam parameters at least equal to those at the outlet from the turbines in a heat and power plant. Water boilers with an adequate heat output can serve to cover the heating load Q_h in the substituting heating plant.

The electrical power P_{CP} at the generator terminals in the substituting condensing power plant amounts to

$$P_{CP} = \delta_P P_{CH} \tag{4.17}$$

where:

P_{CH} – electrical power at the generator terminals in the heat and power plant, equal to the power output P_b of the back-pressure heat and power plant, or the sum of power outputs $P_b + P_c$ of an extraction-condensing heat and power plant;

δ_P – coefficient embracing different power consumptions for auxiliary power demand in a heat and power plant and in a substituting power plant, power transmission losses from the power plant to the distribution sub-station in the heat and power plant and necessary power reserve in the condensing power plant.

Heat output Q_{HP}, delivered from the substituting heating plant, amounts to

$$Q_{HP} = \delta_Q \, Q_{CH} \qquad (4.18)$$

where:

Q_{CH} – heat output delivered from the heat and power plant, equal to heat output Q_n from the formula (4.7) when delivering steam from the turbine outlets and from the reducing-cooling valve, or the sum of heat outputs $Q_b \, \eta_{he} + Q_{wb}$, if the water boiler takes over the peak-load;

δ_Q – coefficient which takes into account heat transmission losses from the heat and power plant to the distribution sub-station, also possibly reserve heat output in the heating plant.

The instantaneous fuel consumption for the generation of electrical power in a substituting power plant and heat output in a substituting heating plant amounts, respectively, to

$$B_{CP} = P_{CP} \, b_{CP} = \frac{P_{CP}}{H_f \, \eta_{CP}} \qquad (4.19)$$

$$B_{HP} = \frac{Q_{HP}}{H_f \, \eta_{HP}} \qquad (4.20)$$

where:

b_{CP} – specific consumption of equivalent coal in the condensing power plant, related to the power output at the terminals;

η_{CP} – total gross efficiency of a condensing power plant;

η_{HP} – efficiency of a substituting heating plant.

4.1.4 Differences in instantaneous and annual fuel consumption

The combined instantaneous fuel consumption B_{se} in a separate system is the sum of the fuel consumption in both substituting plants

$$B_{se} = B_{CP} + B_{HP} \qquad (4.21)$$

and the combined instantaneous fuel consumption $B_{ch} = B_{CH}$ in a combined system consists of the fuel consumption by the steam and water boilers in a heat and power plant

$$B_{ch} = B_{sb} + B_{wb} \qquad (4.22)$$

Hence the difference between the instantaneous fuel consumption in favour of the combined system amounts to

$$\Delta B = B_{se} - B_{ch} = B_{CP} + B_{HP} - B_{CH} \tag{4.23}$$

The values of instantaneous fuel consumption are usually determined for the peak heat load (peak-load output) of a heat and power plant. From this, we can pass on to the annual fuel consumption where, to determine the annual production of energy and the corresponding fuel consumption, the annual utilisation times are introduced in accordance with the following relationships:

– for electrical energy

$$E_b = P_{bp} T_{bp} \tag{4.24}$$

$$E_c = P_{cp} T_{cp} \tag{4.25}$$

– for heat energy

$$H_b = Q_{bp} T_{bp} \tag{4.26}$$

$$H_r = Q_{rp} T_{rp} \tag{4.27}$$

$$H_{wb} = Q_{wbp} T_{wp} \tag{4.28}$$

where the indices b and c refer to electrical power – back-pressure and condensing, the indices b, r and wb refer correspondingly to the heat output delivered by the turbine, reducing-cooling valve and peak-load water boilers, and the index p indicates the peak-load value of the corresponding electrical power P or heat output Q.

In view of this, the annual consumption of fuel by the steam and water boilers in the heat and power plant amounts to

$$F_{CH} = F_{sb} + F_{wb} \tag{4.29}$$

only that, in a back-pressure heat and power plant

$$F_{sb} = F_e + F_h \tag{4.30}$$

and in an extraction-condensing heat and power plant

$$F_{sb} = F_{eb} + F_{ec} + F_h \tag{4.31}$$

where all values of F with appropriate indices are calculated on the basis of the formulae given, concerning the instantaneous consumption of fuel B, but instead of electrical power P_b and P_c and heat output Q_b, Q_r and Q_{wb}, the annual production of electrical energy E_b and E_c, also heat energy H_b, H_r and H_{wb} are assumed correspondingly, and the efficiencies given in these formulae are treated as mean annual efficiencies.

The annual fuel consumption in substituting plants, on the other hand, amounts to:

in a substituting condensing power plant

$$F_{CP} = \frac{E_{CP}}{H_f \eta_{aCP}} \tag{4.32}$$

where

$$E_{CP} = \delta_E \, E_{CH} = \delta_E \, (E_b + E_c) \qquad (4.33)$$

in a substituting heating plant

$$F_{HP} = \frac{H_{HP}}{H_f \, \eta_{aHP}} \qquad (4.34)$$

where

$$H_{HP} = \delta_H \, H_{CH} = \delta_H \, (H_b + H_r + H_{wb}) \qquad (4.35)$$

where:

δ_E, δ_H – conversion coefficients for electrical and heat energy, being the corresponding of conversion coefficients for power δ_P, δ_Q from the formulae (4.17) and (4.18);

η_{aCP}, η_{aHP} – mean annual efficiencies of a condensing power plant and a heating plant, respectively.

The combined annual fuel consumption F_{se} in a separate system is the sum of the corresponding fuel consumption in both substituting plants

$$F_{se} = F_{CP} + F_{HP} \qquad (4.36)$$

and the combined fuel consumption in a combined production system $F_{ch} = F_{CH}$. Hence the difference in the annual fuel consumption in favour of a combined production system amounts to

$$\Delta F = F_{se} - F_{ch} = F_{CP} + F_{HP} - F_{CH} \qquad (4.37)$$

Example 4.1

Calculate the annual saving in fuel consumption in a heat and power plant, as compared with an equivalent separate system. The heat and power plant consists of four identical extraction-condensing turbogenerator sets with a gross electrical power of $P_{Tr} = 25$ MW and gross heat output $Q_{br} = 60$ MJ/s, from Examples 1.3 and 1.6 fed from four steam boilers of appropriate capacity and of four peak-load water boilers with a heat output of $Q_{wbr} = 60$ MJ/s and efficiency $\eta_{wb} = 0.75$.

Assuming that the heat and power plant delivers a heat output to cover the heating demands, the following heating and electrical peak-load utilisation times are assumed:

back-pressure electrical power and heat output
$T_{bp} = 3000$ h/a,
condensing electrical power $T_{cp} = 4000$ h/a,
peak-load water boilers $T_{wp} = 1000$ h/a.

In an equivalent separate system the mean annual efficiencies $\eta_{aCP} = 0.30$; $\eta_{aHP} = 0.70$ and conversion coefficients $\delta_E = 1.1$; $\delta_H = 1.0$ have been assumed.

Solution:
From the Examples 1.3 and 1.6 the parameters of steam and efficiency of the heat and power plant, amount to:
$p_0 = 8\cdot8$ MPa; $t_0 = 535$ °C; $p_e = 0\cdot25$ MPa; $p_c = 5$ kPa; $\eta_{sb} = 0\cdot80$; $\eta_{pi} = 0\cdot99$; $\eta_{cc} = 0\cdot337$; $\eta_{ih} = 0\cdot75$; $\eta_{il} = 0\cdot80$; $\eta_{em} = 0\cdot94$.

In Example 1.3 a back-pressure electrical power output at the generator terminals of $P_b = 16$ MW and a corresponding condensing power output of $P_c = 9$ MW were calculated.

The partial efficiencies of generating energy in a heat and power plant, calculated from the formulae (4.13), (4.14) and (4.6) amount to:

$$\eta_{eb} = \eta_{sb}\,\eta_{pi}\,\eta_{em} = 0\cdot80\cdot0\cdot99\cdot0\cdot94 = 0\cdot745$$
$$\eta_{ec} = \eta_{sb}\,\eta_{pi}\,\eta_{cc}\,\eta_{em} = 0\cdot80\cdot0\cdot99\cdot0\cdot337\cdot0\cdot94 = 0\cdot251$$
$$\eta_h = \eta_{sb}\,\eta_{pi} = 0\cdot80\cdot0\cdot99 = 0\cdot792$$

The annual production of electrical and heat energy in a heat and power plant is calculated from the formulae (4·24–4·28) assuming that the peak-load is equal to the rated output of each of the turbine sets and water boilers, respectively

$$E_b = P_{bp}\,T_{bp} = 4\cdot16\cdot3000\cdot10^{-3} = 192 \text{ GWh/a}$$
$$E_c = P_{cp}\,T_{cp} = 4\cdot9\cdot4000\cdot10^{-3} = 144 \text{ GWh/a}$$
$$H_b = Q_{bp}\,T_{bp} = 4\cdot60\cdot3000\cdot3600\cdot10^{-6} = 2592 \text{ TJ/a}$$
$$H_{wb} = Q_{wbp}\,T_{wp} = 4\cdot60\cdot1000\cdot3600\cdot10^{-6} = 864 \text{ TJ/a}$$

The annual consumption of equivalent coal by the steam and water boilers in a heat and power plant amounts to the following, from the formulae (4.29) and (4.31):

$$F_{eb} = \frac{E_b}{H_f\,\eta_{eb}} = \frac{192\cdot3600}{29\cdot3\cdot0\cdot745} = 31\cdot7\cdot10^3 \text{ t/a}$$

$$F_{ec} = \frac{E_c}{H_f\,\eta_{ec}} = \frac{144\cdot3600}{29\cdot3\cdot0\cdot251} = 70\cdot5\cdot10^3 \text{ t/a}$$

$$F_h = \frac{H_b}{H_f\,\eta_h} = \frac{2592\cdot10^3}{29\cdot3\cdot0\cdot792} = 111\cdot7\cdot10^3 \text{ t/a}$$

$$F_{sb} = F_{eb} + F_{ec} + F_h = 213\cdot9\cdot10^3 \text{ t/a}$$

$$F_{wb} = \frac{H_{wb}}{H_{f\,wb}} = \frac{864\cdot10^3}{29\cdot3\cdot0\cdot75} = 39\cdot3\cdot10^3 \text{ t/a}$$

$$F_{ch} = F_{CH} = F_{sb} + F_{wb} = 253\cdot10^3 \text{ t/a}$$

The annual production of electrical and heat energy in substituting plants, using formulae (4.33) and (4.35) amounts to:

$$E_{CP} = \delta_E(E_b + E_c) = 1{\cdot}1 \cdot 336 = 370 \text{ GWh/a}$$
$$H_{HP} = \delta_H(H_b + H_{wb}) = 1{\cdot}0 \cdot 3456 = 3456 \text{ TJ/a}$$

The annual consumption of equivalent coal in substituting plants, using formulae (4.32), (4.34) and (4.36), amounts to:

$$F_{CP} = \frac{E_{CP}}{H_f \eta_{aCP}} = \frac{370 \cdot 3600}{29{\cdot}3 \cdot 0{\cdot}30} = 151{\cdot}5 \cdot 10^3 \text{ t/a}$$

$$F_{HP} = \frac{H_{HP}}{H_f \eta_{aHP}} = \frac{3456 \cdot 10^3}{29{\cdot}3 \cdot 0{\cdot}70} = 168{\cdot}5 \cdot 10^3 \text{ t/a}$$

$$F_{se} = F_{CP} + F_{HP} = 320 \cdot 10^3 \text{ t/a}$$

The difference in the consumption of equivalent coal in favour of a combined production system, according to the formula (4.37)

$$\Delta F = F_{se} - F_{ch} = (320 - 253) \cdot 10^3 = 67 \cdot 10^3 \text{ t/a}$$

4.2 Costs of generating energy in combined systems

4.2.1 Investment costs in heat and power plants

In conceptual and design works, the investment costs in heat and power plants can be determined on the basis of the indices of specific costs in individual component parts of such a plant: high-pressure steam boilers, turbogenerator sets, and hot water boilers. The total investment costs I can be calculated from the model relationships

$$I = I_{sb} + I_t + I_{wb} \tag{4.38}$$

$$I_{sb} = k_{sb} N_{sb} D_{sbr} \tag{4.39}$$

$$I_t = k_t N_t P_{Tr} \tag{4.40}$$

$$I_{wb} = k_{wb} N_{wb} Q_{wbr} \tag{4.41}$$

in which:
I_{sb} – partial investment costs falling to the high-pressure steam boilers and general needs;
I_t – partial investment costs falling to the turbogenerator sets and electrical equipment;
I_{wb} – partial investment costs falling to the water boilers;
N_{sb} – number of steam boilers;
D_{sbr} – rated capacity of the steam boiler;
N_t – number of turbogenerator sets;

P_{Tr} – rated electrical power output of the turbogenerator set at the generator terminals;
N_{wb} – number of water boilers;
Q_{wbr} – rated heat output of the water boiler.

Specific investment costs can be determined on the basis of tenders or estimates. In study and design works, they can be calculated in the following way:
for the high-pressure steam boilers

$$k_{sb} = a_b \, c_b \, c_u \, (D_{sbr})^{-w_b} \tag{4.42}$$

for the turbogenerator sets

$$k_t = a_t \, c_t \, c_u \, (P_{Tr})^{-w_t} \tag{4.43}$$

for the hot water boilers

$$k_{wb} = c_w \, c_u \, (Q_{wbr})^{-w_w} \tag{4.44}$$

where:
k_{sb} – specific investment costs for the steam boilers, m.u./(t · h^{-1})*;
k_t – specific investment costs for the turbogenerator sets, m.u./MW;
k_{wb} – specific investment costs for the hot water boilers, m.u./(MJ · s^{-1});
a_b, a_t – coefficients dependent on the parameters of live steam;
c_b, c_t, c_w – indices dependent of the price level;
w_b, w_t, w_w – exponentials;
c_u – coefficient dependent on the number of units, which can be calculated from the formula

$$c_u = 0.25N^{-0.3} + 0.75 \tag{4.45}$$

where N – the number of units (steam or water boilers or turbogenerator sets).

Fig. 4.1 presents the dependence of specific investments k_{sb} for the steam boilers on the number and capacity of boilers, and Fig. 4.2 the dependence of the specific investments k_t for the turbogenerator sets on the number and output of back-pressure units. Both graphs were drawn up assuming that $a_b = a_t = 1$ and with inlet parameters of 3.4 MPa, 435 °C. At higher inlet parameters, the following coefficients can be assumed:

6·4 MPa, 465 ÷ 485 °C: $a_b = 1·10 \div 1·13$; $a_t = 1·03 \div 1·05$
8·8 MPa, 500 ÷ 535 °C: $a_b = 1·30 \div 1·40$; $a_t = 1·12 \div 1·20$
12·5 MPa, 535 ÷ 565 °C: $a_b = 1·50 \div 1·65$; $a_t = 1·25 \div 1·35$

Fig. 4.3 presents the dependence of specific investments k_{wb} for the hot water boilers on the number and heat output of boilers.

---

* m.u. = monetary unit.

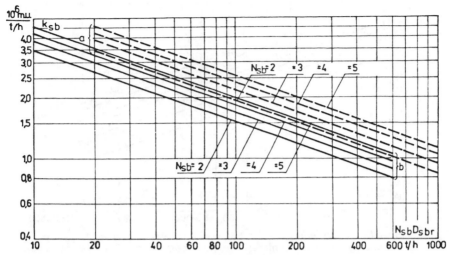

Fig. 4.1   *Specific investment costs for high-pressure steam boilers in a heat and power plant depending upon the number and capacity of boilers ($a_b = 1·0$; $w_b = 0·35$): (a) pulverised fuel boilers ($c_b = 8·0·10^6$); (b) stoker-fired boilers ($c_b = 6·2·10^6$)*

Example 4.2

Calculate the total investment costs in the heat and power plant from Example 4.1, fitted with four pulverised fuel steam boilers with a rated capacity of $D_{sbr} = 140$ t/h, four extraction-condensing turbogenerator sets with inlet parameters $p_0 = 8·8$ MPa, $t_0 = 535$ °C, with a gross electrical power output of $P_{Tr} = 25$ MW and heat output $Q_{br} = 60$ MJ/s and four pulverised fuel water boilers with a heat output of $Q_{wbr} = 60$ MJ/s.

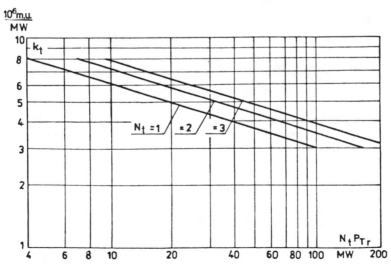

Fig. 4.2   *Specific investment costs for back-pressure turbogenerator sets in a heat and power plant depending upon the number and capacity of back-pressure turbogenerators ($a_t = 1·0$; $c_t = 12·10^6$; $w_t = 0·3$)*

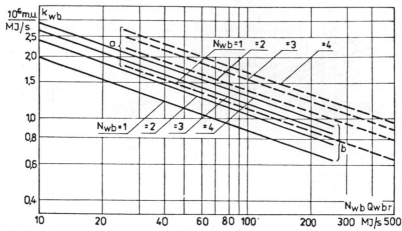

Fig 4.3   *Specific investment costs for hot water boilers in a heat and power plant depending upon the number and heat output of boilers (w_w = 0·35): (a) pulverised fuel boilers (c_w = 5·6· 10^6); (b) stoker-fired boilers (c_w = 4·3· 10^6)*

Solution:
The specific investment costs for the high-pressure steam boilers are calculated from the formula (4.42) or by means of the graphs from Fig. 4.1, assuming the following coefficient values:

a_b = 1·4 for inlet parameters 8·8 MPa, 535 °C,
c_b = 8 · 10^6 for pulverised fuel steam boilers,
w_b = 0·35 for all steam boilers,
c_u = 0·915 from the formula (4·45) at N_{sb} = 4.

Hence
$$k_{sb} = a_b \, c_b \, c_u (D_{sbr})^{-w_b} = 1·4 \cdot 8 \cdot 10^6 \cdot 0·915 \cdot 140^{-0·35}$$
$$= 1·81 \cdot 10^6 \text{ m.u.}/(t \cdot h^{-1})$$

Specific investment costs for the turbogenerator sets are calculated from the formula (4.43) or by means of the graphs from Fig. 4.2, assuming the following coefficient values:

a_t = 1·2 for inlet parameters 8·8 MPa, 535 °C,
c_t = 12 · 10^6 for back-pressure turbine sets,
w_t = 0·3 for all turbine sets,
c_u = 0·915 from the formula (4·45) at N_t = 4.

Hence
$$k_t = a_t \, c_t \, c_u (P_{Tr})^{-w_t} = 1·2 \cdot 12 \cdot 10^6 \cdot 0·915 \cdot 25^{-0·3}$$
$$= 5·0 \cdot 10^6 \text{ m.u./MW}$$

Specific investment costs for the hot water boilers are calculated from the formula (4·44) or by means of the graphs from Fig. 4.3, assuming the following coefficient values:

$c_w = 5\cdot6 \cdot 10^6$ for pulverised fuel water boilers,

$w_w = 0\cdot35$ for all water boilers,

$c_u = 0\cdot915$ from the formula (4·45) at $N_{wb} = 4$.

Hence

$$k_{wb} = c_w\,c_u(Q_{wbr})^{-w_w} = 5\cdot6 \cdot 10^6 \cdot 0\cdot915 \cdot 60^{-0\cdot35}$$
$$= 1\cdot22 \cdot 10^6\ \text{m.u.}/(\text{MJ}\cdot\text{s}^{-1})$$

The total investment costs in the heat and power plant, according to formula (4·38), amount to

$$I = I_{sb} + I_t + I_{wb} = k_{sb}N_{sb}D_{sbr} + k_t N_t P_{Tr} + k_{wb}N_{wb}Q_{wbr}$$
$$= 1\cdot81 \cdot 10^6 \cdot 4 \cdot 140 + 5\cdot0 \cdot 10^6 \cdot 4 \cdot 25 + 1\cdot22 \cdot 10^6 \cdot 4 \cdot 60$$
$$= 1800 \cdot 10^6\ \text{m.u.}$$

4.2.2. Annual costs of heat and power generation

The annual costs K of the combined generation of electrical and heat energy in a heat and power plant, similar to those of the separate generation of the two kinds of energy in an equivalent separate system, embrace the components

$$K = K_r + K_o \tag{4.46}$$

$$K_r = r\,K_{id} = \frac{p(1+p)^N}{(1+p)^N - 1}\,K_{id} \tag{4.47}$$

$$K_o = K_{oc} + K_{ov} \tag{4.48}$$

$$K_{oc} = r_o\,K_i \tag{4.49}$$

$$K_{ov} = c_f\,F \tag{4.50}$$

$$K_{id} = k_z\,K_i \tag{4.51}$$

where:

K_r – annual reproduction costs (costs of amortisation and accumulation), m.u./a;

K_o – annual operating costs, m.u./a;

K_{oc} – annual fixed operating costs, m.u./a;

K_{ov} – annual variable operating costs, m.u./a;

r – annual reproduction rate;

p – accumulation rate;

N – number of operation years;

r_o – coefficient of annual fixed operating costs, covering costs of general overhauls and running costs, maintenance costs including social security and administrative costs;

k_z – coefficient of freezing investment costs;
$K_i = I$ – total investment costs without freezing, m.u.;
K_{id} – discounted investment costs taking into account freezing during the building period, m.u.;
c_f – price of equivalent coal, m.u./t;
F – annual consumption of equivalent coal, t/a.

Substituting the operating costs from (4.48) into (4.46), the total annual costs are obtained, as the sum of three components

$$K = K_r + K_{oc} + K_{ov} \qquad (4.52)$$

The annual costs can also be presented as the sum of annual fixed K_c and variable K_v costs

$$K = K_c + K_v \qquad (4.53)$$
$$\text{here} \quad K_c = K_r + K_{oc} \qquad (4.54)$$
$$K_v = K_{ov} \qquad (4.55)$$

Calculating in turn the annual costs K for the heat and power plant considered (CH), the substituting condensing power plant (CP) and substituting heating plant (HP), the difference in annual costs ΔK in favour of a combined system can be determined

$$\Delta K = K_{CP} + K_{HP} - K_{CH} \qquad (4.56)$$

The condition for the profitability of combined production is therefore $\Delta K > 0$.

Example 4.3
Calculate the overall annual costs for the heat and power plant from Examples 4.1 and 4.2 and compare them with the annual costs for separate generation of electrical energy in a substituting condensing power plant and heat energy in a substituting heating plant, assuming that:

- for all plants considered:
  accumulation rate $p = 0 \cdot 08$
  number of operation years $N = 25$
  price of equivalent coal $c_f = 700$ m.u./t
- for particular plants:
  coefficient of investment freezing:
  in the heat and power plant $k_{zCH} = 1 \cdot 2$
  in the condensing power plant $k_{zCP} = 1 \cdot 2$
  in the heating plant $k_{zHP} = 1 \cdot 1$
  coefficient of annual fixed operating costs:
  in the heat and power plant $r_{oCH} = 0 \cdot 05$
  in the condensing power plant $r_{oCP} = 0 \cdot 05$
  in the heating plant $r_{oHP} = 0 \cdot 04$

specific investment costs:
in the condensing power plant $k_{iCP} = 11 \cdot 10^6 \,\mathrm{m.u./MW}$
in the heating plant $k_{iHP} = 1 \cdot 3 \cdot 10^6 \,\mathrm{m.u./(MJ \cdot s^{-1})}$

Solution:
In the heat and power plant, the annual costs calculated on the basis of the formulae (4.46) − (4.51) cover:
costs of amortisation and accumulation for the investment costs from Example 4.2

$$K_{rCH} = r K_{idCH} = \frac{p(1+p)^N}{(1+p)^N - 1} k_{zCH} K_{iCH}$$

$$= \frac{0 \cdot 08 \cdot 1 \cdot 08^{25}}{1 \cdot 08^{25} - 1} \cdot 1 \cdot 2 \cdot 1800 \cdot 10^6 = 202 \cdot 3 \cdot 10^6 \,\mathrm{m.u./a}$$

annual fixed operating costs

$$K_{ocCH} = r_{oCH} K_{iCH} = 0 \cdot 05 \cdot 1800 \cdot 10^6 = 90 \cdot 0 \cdot 10^6 \,\mathrm{m.u./a}$$

annual variable operating costs of fuel consumption from Example 4.1

$$K_{ovCH} = c_f F_{CH} = 700 \cdot 253 \cdot 10^3 = 177 \cdot 1 \cdot 10^6 \,\mathrm{m.u./a}$$

Hence the overall annual costs for the heat and power plant
$$K_{CH} = K_{rCH} + K_{ocCH} + K_{ovCH} = (202 \cdot 3 + 90 \cdot 0 + 177 \cdot 1) \cdot 10^6$$
$$= 469 \cdot 4 \cdot 10^6 \,\mathrm{m.u./a}$$

The annual costs in a substituting condensing power plant embrace:
the costs of amortisation and accumulation for the gross electrical power, amounting to

$$P_{CP} = \delta_P N_t P_{Tr} = 1 \cdot 1 \cdot 4 \cdot 25 = 110 \,\mathrm{MW}$$
$$K_{rCP} = r k_{zCP} k_{iCP} P_{CP} = 0 \cdot 09368 \cdot 1 \cdot 2 \cdot 11 \cdot 10^6 \cdot 110$$
$$= 136 \cdot 0 \cdot 10^6 \,\mathrm{m.u./a}$$

annual fixed operating costs

$$K_{ocCP} = r_{oCP} k_{iCP} P_{CP} = 0 \cdot 05 \cdot 11 \cdot 10^6 \cdot 110$$
$$= 60 \cdot 5 \cdot 10^6 \,\mathrm{m.u./a}$$

annual variable operating costs of fuel consumption from Example 4.1

$$K_{ovCP} = c_f F_{CP} = 700 \cdot 151 \cdot 5 \cdot 10^3 = 106 \cdot 1 \cdot 10^6 \,\mathrm{m.u./a}$$

Hence the overall annual costs for a condensing power plant

$$K_{CP} = K_{rCP} + K_{ocCP} + K_{ovCP} = (136 \cdot 0 + 60 \cdot 5 + 106 \cdot 1) \cdot 10^6$$
$$= 302 \cdot 6 \cdot 10^6 \,\mathrm{m.u./a}$$

The annual costs in a substituting heating plant embrace: the costs of amortisation and accumulation for a gross heat output, amounting to

$$Q_{HP} = \delta_Q(N_t Q_{br} + N_{wb} Q_{wbr}) = 1 \cdot 0 \, (4 \cdot 60 + 4 \cdot 60) = 480 \text{ MJ/s}$$

$$K_{rHP} = r \, k_{zHP} \, k_{iHP} \, Q_{HP} = 0 \cdot 09368 \cdot 1 \cdot 1 \cdot 1 \cdot 3 \cdot 10^6 \cdot 480$$
$$= 64 \cdot 3 \cdot 10^6 \text{ m.u./a}$$

annual fixed operating costs

$$K_{ocHP} = r_{oHP} \, k_{iHP} \, Q_{HP} = 0 \cdot 04 \cdot 1 \cdot 3 \cdot 10^6 \cdot 480$$
$$= 25 \cdot 0 \cdot 10^6 \text{ m.u./a}$$

annual variable operating costs of fuel consumption from Example 4.1

$$K_{ovHP} = c_f \, F_{HP} = 700 \cdot 168 \cdot 5 \cdot 10^3 = 118 \cdot 0 \cdot 10^6 \text{ m.u./a}$$

Hence the overall annual costs for a heating plant

$$K_{HP} = K_{rHP} + K_{ocHP} + K_{ovHP} = (64 \cdot 3 + 25 \cdot 0 + 118 \cdot 0) \cdot 10^6$$
$$= 207 \cdot 3 \cdot 10^6 \text{ m.u./a}$$

The difference in annual costs in favour of the combined system, from the formula (4·56) amounts to

$$\Delta K = K_{CP} + K_{HP} - K_{CH} = (302 \cdot 6 + 207 \cdot 3 - 469 \cdot 4) \cdot 10^6$$
$$= 40 \cdot 5 \cdot 10^6 \text{ m.u./a}$$

In order to determine the calculated annual costs which would be equivalent to the annual costs which change in the successive years of operation of the plant, the characteristics, or distribution in time, of the following costs and effects should be established:

investment costs in each year of construction of the plant;
operating costs in each year the plant exists;
production effects, i.e. heat or electrical energy generated in successive years.

The time characteristic of the investment costs has a general form

Year	$-M$	...	$-j$	...	-2	-1	0
Investments	$K_{i(-M)}$	...	$K_{i(-j)}$	...	$K_{i(-2)}$	$K_{i(-1)}$	K_{i0}

Investment costs reduced to the zero year by discount method, are calculated from the formula

$$K_{id} = \sum_{j=0}^{M} K_{i(-j)}(1 + p)^j \qquad (4.57)$$

where:
p – the discount rate equal to the assumed rate of accumulation;
here the costs distributed over the years from $(-M)$ to 0 are 'frozen' investments during the period the object is under construction.

The time characteristic of annual operating costs has a general form

Year	0	1	2	...	i	...	N
Operating cost	–	K_{o1}	K_{o2}	...	K_{oi}	...	K_{oN}

For this time characteristic of costs, it is possible to calculate by discount method:

– either the sum of costs discounted for the zero year

$$K_{od} = \sum_{i=1}^{N} K_{oi}(1+p)^{-i} \tag{4.58}$$

– or the sum of costs discounted for the Nth year

$$K'_{od} = \sum_{i=1}^{N} K_{oi}(1+p)^{N-i} = K_{od}(1+p)^N \tag{4.59}$$

On this basis it is possible to calculate the mean discounted annual operating cost

$$K_{oa} = r\, K_{od} = \frac{p(1+p)^N}{(1+p)^N - 1} \sum_{i=1}^{N} K_{oi}(1+p)^{-i} \tag{4.60}$$

or

$$K_{oa} = a_{am} K'_{od} = \frac{p}{(1+p)^N - 1} \sum_{i=1}^{N} K_{oi}(1+p)^{N-i} \tag{4.61}$$

In this way, the complex time characteristic of investment and operating costs is substituted by a simple characteristic in which the investment costs K_{id} are concentrated in the zero year and the operating costs K_{oa} are the same each year $i = 1, 2, \ldots, N$. For the simple time characteristic the equivalent annual costs amount to

$$K = r\, K_{id} + K_{oa} \tag{4.62}$$

The time characteristic of production effects, i.e. the distribution of production (e.g. heat energy) in consecutive years, has a general form

Year	0	1	2	...	i	...	N
Annual effect	–	H_1	H_2	...	H_i	...	H_N

For this time characteristic of effects it is possible to calculate by discount method:

– either the sum of effects, discounted for the zero year

$$H_{d0} = \sum_{i=1}^{N} H_i(1+p)^{-i} \tag{4.63}$$

– or the sum of effects, discounted for the Nth year

$$H'_{d0} = \sum_{i=1}^{N} H_i(1+p)^{N-i} \tag{4.64}$$

On this basis it is possible to calculate the mean discounted annual effect

$$H_a = r H_{d0} = \frac{p(1+p)^N}{(1+p)^N - 1} \sum_{i=1}^{N} H_i(1+p)^{-i} \tag{4.65}$$

or

$$H_a = a_{am} H'_{d0} = \frac{p}{(1+p)^N - 1} \sum_{i=1}^{N} H_i(1+p)^{N-i} \tag{4.66}$$

Table 4.1   *Discounting coefficients for different values of N for p = 0·08*

N	$(1+p)^N$	$(1+p)^{-N}$	a_{am}	r
1	1·080 00	0·925 93	1·000 00	1·080 00
2	1·166 40	0·857 34	0·480 77	0·560 77
3	1·259 71	0·793 83	0·308 03	0·388 03
4	1·360 49	0·735 03	0·221 92	0·301 92
5	1·469 33	0·680 58	0·170 46	0·250 46
6	1·586 87	0·630 17	0·136 32	0·216 32
7	1·713 82	0·583 49	0·112 07	0·192 07
8	1·850 93	0·540 27	0·094 01	0·174 01
9	1·999 00	0·500 25	0·080 08	0·160 08
10	2·158 92	0·463 19	0·069 03	0·149 03
11	2·331 64	0·428 88	0·060 08	0·140 08
12	2·518 17	0·397 11	0·052 70	0·132 70
13	2·719 62	0·367 70	0·046 52	0·126 52
14	2·937 19	0·340 46	0·041 30	0·121 30
15	3·172 17	0·315 24	0·036 83	0·116 83
16	3·425 94	0·291 89	0·032 98	0·112 98
17	3·700 02	0·270 27	0·029 63	0·109 63
18	3·996 02	0·250 25	0·026 70	0·106 70
19	4·315 70	0·231 71	0·024 13	0·104 13
20	4·660 96	0·214 55	0·021 85	0·101 85
25	6·848 48	0·146 02	0·013 68	0·093 68
30	10·062 66	0·099 38	0·008 83	0·088 83
35	14·785 34	0·067 63	0·005 80	0·085 80
40	21·724 52	0·046 03	0·003 68	0·083 68
45	31·920 45	0·031 33	0·002 59	0·082 59
50	46·901 61	0·021 32	0·001 74	0·081 74

Table 4.1 gives the values of the coefficients a_{am}, r, $(1+p)^N$ and $(1+p)^{-N}$ for $p = 0·08$.

Example 4.4
The investment costs calculated in Example 4.2 are distributed equally over four years – the period the heat and power plant is under construction – and amount to 450 million monetary units (m.u.) each year. Calculate the equivalent costs reduced to the zero year at a discount rate of $p = 0.08$ and the coefficient of 'freezing' the costs during building.

Solution
Each year $(-j)$ the investment costs amount to

$$K_{i(-j)} = 450 \cdot 10^6 \text{ m.u.}; \quad j = 0, 1, 2, 3$$

In view of this, the sum of investments reduced to the zero year (discounted sum), according to the formula (4.57) amounts to

$$k_{id} = \sum_{j=0}^{3} K_{i(-j)}(1+p)^j$$
$$= 450 \cdot 10^6 (1 \cdot 08^0 + 1 \cdot 08^1 + 1 \cdot 08^2 + 1 \cdot 08^3)$$
$$= 2027 \cdot 7 \cdot 10^6 \text{ m.u.}$$

and the coefficient of 'freezing', defined as the ratio of the discounted total to the arithmetic total of investments amounts to

$$k_z = \frac{K_{id}}{K_i} = \frac{2027 \cdot 7 \cdot 10^6}{4 \cdot 450 \cdot 10^6} = 1 \cdot 1265$$

Example 4.5
Operating costs of energy differ each year. During the first ten years the increase in costs is due to increased load, in the 11th year there is a drop in costs due to the commissioning of a second, parallel transmission line, and in following years, to the end of the operating period lasting $N = 20$ years, the costs increase again due to a further increase in load.

Calculate the mean discounted annual operating costs, if the rate of discount is $p = 0.08$. A list of operating costs and calculations of discounted costs is given in Table 4.2.

Solution:
The sum of operating costs discounted for the 20th year, according to the formula (4.59)

$$K'_{od} = \sum_{i=1}^{N} K_{oi}(1+p)^{N-i}$$
$$= \sum_{i=1}^{20} K_{oi} \cdot 1 \cdot 08^{20-i} = 340 \cdot 6 \cdot 10^6 \text{ m.u}$$

Table 4.2   *Discounted operating costs in Example 4.5*

Year i	Operating costs K_{oi}	Discounting factor $(1+p)^{N-i}$	Discounted costs $K_{oi}(1+p)^{N-i}$
–	10^6 m.u./a	–	10^6 m.u./a
1	6·00	4·316	25·90
2	6·21	3·996	24·82
3	6·44	3·700	23·83
4	6·69	3·426	22·92
5	6·96	3·172	22·08
6	7·25	2·937	21·29
7	7·56	2·720	20·56
8	7·89	2·518	19·87
9	8·26	2·332	19·26
10	8·61	2·159	18·59
11	7·50	1·999	14·99
12	7·70	1·851	14·25
13	7·92	1·714	13·57
14	8·14	1·587	12·92
15	8·38	1·469	12·31
16	8·62	1·360	11·72
17	8·88	1·260	11·19
18	9·14	1·166	10·66
19	9·42	1·080	10·17
20	9·70	1·000	9·70
	Total	K'_{od}	340·60

In view of this, the mean discounted annual operating costs, identical in each year of the operating period, according to the formula (4.61), amount to

$$K_{oa} = \frac{p}{(1+p)^N - 1} K'_{od} = \frac{0·08}{1·08^{20} - 1} 340·6 \cdot 10^6$$

$$= 7·44 \cdot 10^6 \text{ m.u./a}$$

4.3 Influence of heat transmission costs

4.3.1 Economic range of steam transmission
The subject considered in this section is the problem of the economic range of transmitting heat in the form of steam from a back-pressure heat and power

plant supplying several neighbouring industrial plants. It is frequently the case, in practice, that two neighbouring factories build their own heating plants or heat and power plants, instead of an economically justified joint source of heat and electrical energy.

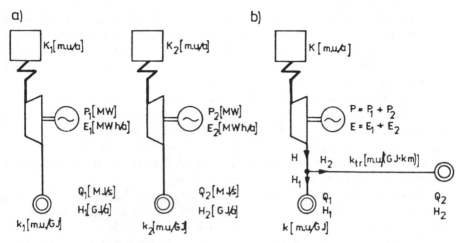

Fig. 4.4   *Diagram of the connecting up of neighbouring industrial plants to a joint heat and power plant*

When determining the economic range of supplying heat energy in this case, two variants should be considered, these being presented in Fig. 4.4:

(a) two separate industrial heat and power plants, each of which supplies its own factory with heat and electrical energy, operating in tandem with the power system;

(b) one larger industrial heat and power plant supplying its own factory with heat and electrical energy and the second factory with heat energy; thus the second factory does not possess its own heat and power plant and draws steam from the first factory and electrical energy from the external grid which is also connected to the first factory.

The distance which should be considered to constitute an economic range of transmitting heat is that at which the annual heat transmission costs are equal to the benefits resulting from lower generating costs in a joint heat and power plant in relation to those in a smaller independent heat and power plant which a given factory would have to build to cover its own requirements.

Comparison of the costs of generating and transmitting heat, was carried out on several typical systems considered to be representative of many cases met in practice. Two levels of pressure of process steam delivered from the heat and power plant were assumed: 0·5 MPa and 0·8 MPa. The six typical types of boilers were those with capacities of 4, 6·5, 10, 16, 32, and 64 t/h.

To simplify matters, only such heat and power systems were considered which included two boilers and one back-pressure turbine. The parameters of live steam in the heat and power plants were taken as 2·4 MPa, 400 °C for the 4·0 and 6·5 t/h boilers; 3·7 MPa, 450 °C for the 10 and 16 t/h, and 6·9 MPa, 475 °C for the 32 and 64 t/h boilers.

Next, two variants of utilisation time of the installed capacity of steam turbines were assumed: 4000 h/a and 6000 h/a, where the turbine capacity installed in these variants constitutes 0·75 and 0·9 of the corresponding rated capacity of the boilers. This corresponds to the average situation existing in two- and three-shift plants.

It was further assumed that the steam transmitted to the neighbouring factory amounts to a maximum of 8, 13 or 20 t/h, which corresponds to the capacities of two 4 t/h, two 6·5 t/h, and two 10 t/h boilers, respectively. Finally, 12 calculation alternatives resulting from two back-pressure levels, two utilisation times and three different quantities of steam transmitted were obtained for each size of heat and power plant.

The overall annual costs of generating energy, including both heat and electrical energy in the alternative back-pressure heat and power plants considered, have been determined by the method described in section 4.2.2.

The annual costs of transmitting heat from one factory for which a larger joint heat and power plant is planned, to a second factory to which steam is transmitted, instead of building a separate smaller heat and power plant, have also been calculated by the same method.

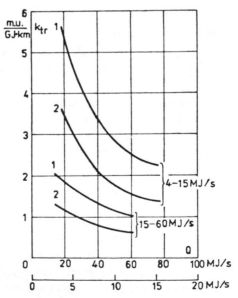

Fig. 4.5   *Specific costs of heat transmission depending upon the heat output transmitted and its utilisation time at a steam pressure of 0·5 MPa (1) T_p = 4000 h/a; (2) T_p = 6000 h/a*

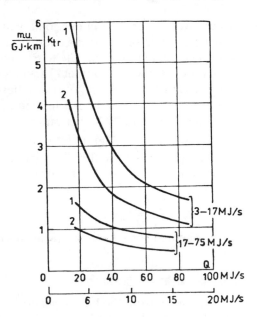

Fig. 4.6 *Specific costs of heat transmission depending upon the heat output transmitted and its utilisation time at a steam pressure of 0·8 MPa (1) T_p = 4000 h/a; (2) T_p = 6000 h/a*

The specific costs of transmitting heat in the form of steam from a heat and power plant, depending upon the heat output transmitted and its utilisation time, are given in Figs. 4.5 and 4.6. In view of the substantial increase in the specific transmission costs with small quantities of heat transmitted, two graphs of heat output variability have been distinguished.

In order to determine the economic range of transmitting heat in the form of steam, if the generation of heat in an industrial heat and power plant is centralised, the total annual costs of heat supplies have been compared in two variants which have been diagrammatically presented in Fig. 4.4.

A. In the case of two individual heat and power plants L km apart (Fig. 4.4a), the joint annual costs include the sum of generating costs $K_1 + K_2$ in the two plants. In the first, electrical energy E_1 and heat energy H_1 are generated, where the corresponding back-pressure power amounts to P_1, and the heat output supplied in back-pressure steam to Q_1. Correspondingly, in the second plant, the energy production amounts to E_2 and H_2, and the power supplied to P_2 and Q_2. The division of costs K_1 or K_2 between electrical and heat energy in each of the plants is a secondary matter, in view of the fact that in combined production, the total annual costs are compared.

B. In the case of centralised generation in one joint heat and power plant situated in the first factory (Fig. 4.4b), its energy production E and H constitutes the corresponding sum of $E_1 + E_2$ and $H_1 + H_2$. The resultant electrical power P and heat output Q are determined similarly. The total annual generating costs K are, of course, less than the sum of $K_1 + K_2$, as both the specific investment costs and operating costs are lower in a plant with a substantially higher output.

In this case, part of the heat energy generated is transmitted by pipeline in the form of steam over distance L from the first factory to the second one, the specific transmission costs amounting to k_{tr} and the total transmission cost – to K_{tr}, where

$$K_{tr} = k_{tr} H_2 L \qquad (4.67)$$

Comparison of the joint annual costs from the two cases described affords the equation

$$K + K_{tr} = K_1 + K_2 \qquad (4.68)$$

from which K_{tr} can be determined from the known generation costs K_1, K_2, K, and from this the limiting distance L of the heat transmission can be found.

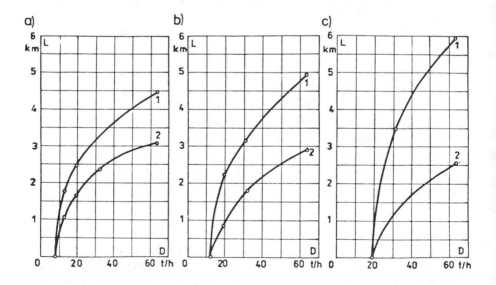

Fig. 4.7   *Economic range of transmitting steam with a pressure of 0·5 MPa at a utilisation time of 4000 h/a:*
(a) transmission of 8 t/h; (b) transmission of 13 t/h; (c) transmission of 20 t/h
1 – limiting distances when extending a heat and power plant to be used by two factories,
2 – limiting distances when utilising the heat output reserve in a heat and power plant serving two factories

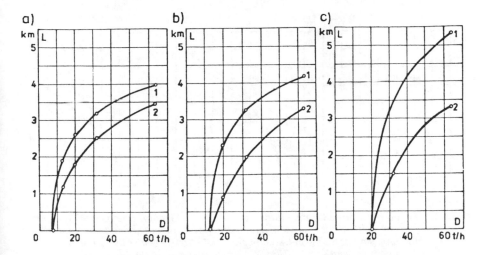

Fig. 4.8 *Economic range of transmitting steam with a pressure of 0·8 MPa at a utilisation time of 4000 h/a (denotation as in Fig. 4.7)*

The limiting distances L calculated by this method are presented in Figs. 4.7 and 4.8, at different quantities and pressures of steam transmitted, at different outputs of the boilers from which the steam is taken, and at a utilisation time of the maximum steam flow of the back-pressure units installed amounting to 4000 h/a (curve 1).

If the quantity of steam transmitted by pipeline to the second factory is much less than that generated in the first factory, or if this transmission is possible without increasing the heat output of the heat and power plant in the first factory, due to the existence in this of a certain reserve of power in relation to local demand, then the centralisation of generation does not result in the reduction of the specific costs in the first factory in relation to the system with two independent heat and power plants.

In such a case, the economic range L' of heat transmission is slightly less than the previously calculated limiting distance L and results in the equation

$$k_{tr} L' = k_2 - k \qquad (4.69)$$

where:
k_2 – specific cost of the individual generation of heat in the second factory;
k – specific cost for the central generation of heat in the first factory.

The economic range L' calculated by this method is presented in Figs. 4.7 and 4.8, depending upon the output of the boilers in the first factory, at various quantities and pressures of steam transmitted, as well as the utilisation time of 4000 h/a (curve 2). Fig. 4.9 presents the results of a similar analysis carried out with an assumed utilisation time of 6000 h/a, at various back-pressures of steam received. To simplify matters, only the method resulting from equation (4.69) was applied in this analysis.

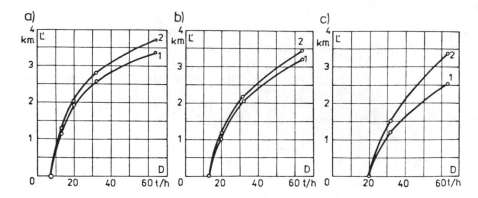

Fig. 4.9 *Economic range of transmitting steam if the utilisation time is 6000 h/a in the case of heat output reserve in a heat and power plant serving two factories: (a) transmission of 8 t/h; (b) transmission of 13 t/h; (c) transmission of 20 t/h*
1 – steam pressure of 0·5 MPa, 2 – steam pressure of 0·8 MPa

The following conclusions derive from the analysis of examples of the economic range of heat transmitted in the form of steam presented:

1. If the heat demand by an industrial plant corresponds to a peak demand of over 60 t/h of steam with a pressure of 0·5 MPa or 0·8 MPa and if the distance from the heat and power plant of the neighbouring factory with an appropriately higher output is over 5 km, the rule is usually that the first factory should design its own source of heat energy. An exception is the case in which the heat supply to a region is satisfied by the construction of a high-pressure steam network (e.g. 0·8 ÷ 1 MPa) supplied from a large district heat and power plant.

2. If an industrial plant has a lower demand for steam, of the order of 8 ÷ 20 t/h, the possibility of covering this demand by transmitting steam with the pressure required, from a neighbouring industrial heat and power plant should be considered. If the distance the steam is to be transmitted does not exceed 2 ÷ 3 km, the centralisation of generation in one heat and power plant for the two factories usually proves expedient as compared with individual production, on condition that increasing the demand for steam by introducing an additional receiver does not necessitate the enlargement of the existing or planned joint heat and power plant.

3. At distances of 3 ÷ 5 km the transmission of steam is possible on conditions that pressure and temperature drops are maintained within permissible limits, which requires separate calculations. This transmission is economically justified in such cases in which, in view of increasing the steam supply with its centralised production, there is a distinct reduction in the specific costs of heat generation in the joint heat and power plant, in a system with steam transmission, as compared with a system with two independent sources of heat.

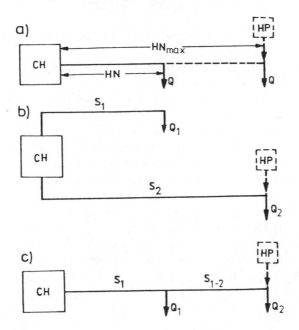

Fig. 4.10   *Typical examples of extending a district heating network supplied from a heat and power plant*

4.3.2 Economic range of hot water district heating network

The subject under consideration in this section is the problem of the economic range of transmitting heat in the form of hot water from a district or industrial heat and power plant via the main heating network (HN). Fig 4.10 presents typical cases of the development of a heat distribution network supplied from a heat and power plant.

The simplest case constitutes one main heating conduit with a supply Q at the terminal (Fig. 4.10a). When extending this main line, the annual cost of transmitting heat in the distribution network K_{HN} increases correspondingly, as at a given heat output Q at the main line terminal, viz. with a specific diameter, this cost is approximately proportional to the length of the main line.

In view of the profitability of the combined generation of heat energy in a heat and power plant (CH), the maximum length of the heat distribution mains is obtained by comparing the sum of the annual costs of generating and transmitting heat with the annual cost of generating heat in a local heating plant (HP)

$$K_{CH} + K_{max\ HN} = K_{HP} \qquad (4.70)$$

where:

K_{CH} – annual cost of generating heat output Q and corresponding heat energy in a heat and power plant;

$K_{max\ HN}$ – the annual cost of transmitting heat output Q in the main heat network over the limitary distance;

K_{HP} – annual cost of generating heat output Q and corresponding heat energy in the local heating plant.

This reasoning is fairly rarely applied, e.g. when considering the admissible distance for locating a heat and power plant away from a town, for economic reasons. One comes across the question of profitability of expanding existing networks much more frequently. Two typical cases, illustrated in Figs. 4.10b and c, can then be considered. Apart from the existing heat consumer Q_1 supplied from the heat and power plant by the main pipeline S_1, a new consumer Q_2 arises, which can be supplied from the new pipeline S_2 (Fig. 4.10b), or by extending the existing one by section S_{1-2} (Fig. 4.10c).

The connecting up of consumer Q_2 to the heat distribution network instead of constructing a new local heating plant at that point is expedient when

$$K_{1+2}^{CH} + K_{1+2}^{HN} < K_{1CH} + K_{1HN} + K_{2CP} + K_{2HP} \qquad (4.71)$$

the indices in this formula denoting:

1 + 2 – combined production embracing both heat consumers Q_1 and Q_2;
1 – combined production covering only the heat consumer Q_1;
2 – separate production for the heat consumer Q_2.

Thus K_{2HP} – annual cost of substitute local heating plant used to supply the consumer Q_2 individually;

K_{2CP} – annual cost of generating – in a substitute condensing power plant – that electrical energy which would be generated on the basis of heat delivered Q_2 in combined production.

The inequality (4.71) brings us to the following expression, embracing the differences in costs

$$\Delta K_{2CH} + \Delta K_{2HN} - K_{2CP} < K_{2HP} \qquad (4.72)$$

where

$$\Delta K_{2CH} = K_{1+2}^{CH} - K_{1CH} \qquad (4.73)$$

$$\Delta K_{2HN} = K_{1+2}^{HN} - K_{1HN} \qquad (4.74)$$

If the inequality (4.72) is fulfilled, it is worth extending the network and connecting up a new consumer Q_2. For practical purposes, the problem of the profitability of extending the heat distribution mains arises most frequently when considering supplying suburban districts in cases similar to those given in Fig. 4.10c. For these cases, one can assume the simplification that the difference in costs of the network ΔK_{2HN} defined by the formula (4.74) constitutes simply the annual cost related to the construction of the main line section S_{1-2}.

The difference between the costs K_{2HP} of the local heating plant for consumer Q_2 and those arising from the formula (4.72)

$$\Delta K_{2CH} - K_{2CP}$$

enables the permissible costs of expanding the heating network ΔK_{2HN} for the section S_{1-2} to be found, according to the relationship

$$(\Delta K_{2HN})_{max} = K_{2HP} - (\Delta K_{2CH} - K_{2CP}) \qquad (4.75)$$

$$(\Delta K_{2HN})_{max} = (K_{2HP} + K_{2CP}) - \Delta K_{2CH} \qquad (4.76)$$

Thus the costs of separate generation K_{2HP} and K_{2CP} should be defined and the difference in the costs ΔK_{2CH} arising in the heat and power plant as the result of connecting up of consumer Q_2, subtracted.

4.4 Limits of combined generation profitability

4.4.1 Methods of investigating the profitability of combined generation systems

Existing methods of investigating the profitability of combined heat and power production in industrial heat and power plants can be divided into two groups:

- the methods in which the costs of electrical energy generated in a back-pressure cycle are compared with the costs of energy which an industrial plant consumes from a condensing power plant;
- the methods in which the costs of heat energy delivered from the back-pressure outlets are compared with the costs of generating heat in a local industrial heating plant.

In this section, the method assumed to investigate the profitability of industrial heat and power plants, in particular to define the lower limit of the range of profitability of combined heat and power production, consists in comparing the heat and power plant investigated with an equivalent power system embracing a condensing power plant and an industrial heating plant. In the compared system, both electrical and heat energy are generated separately, which is defined as separate energy production.

The condition for comparability of combined and separate production is the equality of effects in both systems, which should be technically equivalent in respect of the quantity and quality of electrical and heat energy generated. The equivalent condensing power plant is a component of the electrical power system and is connected up to it by transmission lines and transformer stations. The equivalent heating plant in the compared system is situated on the site of the heat and power plant investigated.

The purpose of investigating the profitability limit of combined heat and power production in industry is to define:

- the lower range of heat outputs, parameters and indices of industrial heating plants, in which the inexpediency of combined heat and power production is so certain as to relieve the designers from the necessity to analyse the problem in detail;
- the intermediate range of heat outputs, parameters and indices at which the expediency of combined heat and power production may depend upon several local conditions and the individual features of the object designed;
- the upper range of heat outputs, parameters and indices at which the expediency of combined heat and power production is, in turn, certain.

The question of profitability of combined production of electrical and heat energy in small industrial heat and power plants, as compared with separate production of heat energy in equivalent heating plants and electrical energy in equivalent power plants, has frequently been analysed in numerous studies. In recent years, however, it was found that in Poland, this limit shifted towards the higher electrical and heat outputs, for various reasons, mainly due to the substantial increase in specific investment costs in heat and power plants.

In the present section, the total annual costs of energy production, calculated from the formula (4.52) have been taken into account to investigate the profitability of combined heat and power production. The questions of reliability of supply and quality of energy have, however, been omitted, although they have an undoubted influence on the profitability of heat and power plants.

4.4.2 Examples of economic calculations for model cases of industrial heat and power plants

In the economic calculation, twelve model heat and power plant systems were chosen [39] and the following values were taken into consideration:

- capacity of steam boilers: $16 \div 70$ t/h;
- number of steam boilers: 2 or 2 + 1 with 1 reserve unit;
- capacity of hot water boilers: $11 \cdot 6 \div 44$ MJ/s;
- total heat output of heat and power plant: $20 \cdot 3 \div 174$ MJ/s;
- utilisation time of peak-load heat output: $2000 \div 6000$ h/a;
- parameters of live steam before the turbine: from $3 \cdot 4$ MPa, 435 °C to $8 \cdot 8$ MPa, 500 °C;
- back-pressure behind the turbine: $0 \cdot 19$ MPa at a solely heating load and $0 \cdot 25 \div 0 \cdot 8$ MPa if the heat load is for process heat.

In twelve equivalent, model industrial heating plants, corresponding steam boilers with capacities of 16 t/h or 32 t/h and outlet parameters of $1 \cdot 5$ MPa, 350 °C together with reducing-cooling valves were adopted, as well as hot water boilers with capacities of $11 \cdot 6 \div 44$ MJ/s, similar to the heat and power plants.

Two alternatives for utilisation times and their corresponding extreme assumptions as to the ratio of heating to process heat load, were adopted, namely, in case A it was assumed that the heat load in the heat and power plant constituted only heating load in hot water, with a utilisation time of 2000 h/a, and in cases B and C – only process steam heat load, with a utilisation time of 6000 h/a.

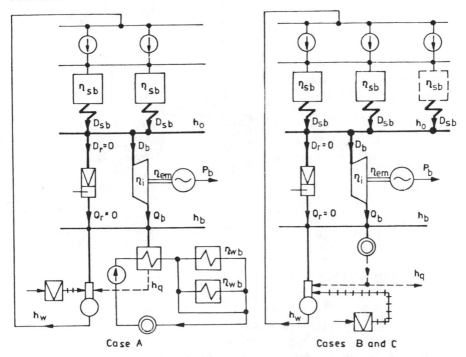

Case A   Cases B and C

Fig. 4.11   *Heat diagrams of model industrial heat and power plants*

Fig. 4.11 presents heat diagrams of model heat and power plants with two steam boilers (cases A and B), or model case C with three steam boilers, one of which being a reserve unit. Fig. 4.12 presents heat diagrams of equivalent model cases A, B and C of industrial heating plants.

System I consists of two steam boilers of 16 t/h each with outlet parameters of 3·7 MPa, 450 °C and one back-pressure turbine with an outlet heat output of 20·3 MJ/s. In case IA, at a peak-load utilisation time of 2000 h/a, two hot water boilers produce an additional peak output of 23·3 MJ/s, as a result of which the total heat output of the heat and power plant amounts to 43·6 MJ/s in hot water. In case IB, at a utilisation time of 6000 h/a, there are no additional hot water boilers in view of which the total heat output of the heat and power plant amounts to 20·3 MJ/s only, which corresponds to the outlet steam flow from a turbine with a capacity of 32 t/h at a back-pressure of 0·25 MPa, 0·5 MPa or 0·8 MPa.

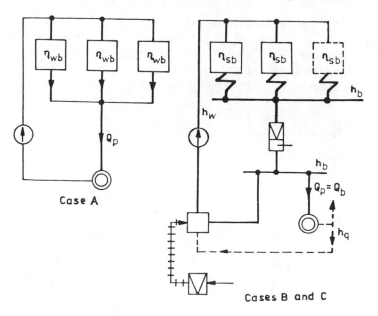

Fig. 4.12   *Heat diagrams of equivalent industrial heating plants*

System II consists of two steam boilers of 32 t/h each with the same outlet parameters as previously 3·7 MPa, 450 °C, and one back-pressure turbine with double the capacity and outlet heat output of 40·7 MJ/s. In case IIA there are two additional hot water boilers which together produce 46·5 MJ/s, thus giving a total heat output of the heat and power plant of 87·2 MJ/s in hot water. In case IIB, without hot water boilers, the total heat output amounts to 40·7 MJ/s in steam of 0·25 MPa, 0·5 MPa or 0·8 MPa.

System III has two steam boilers with the same capacity as in system II, but higher outlet parameters of 6·9 MPa, 500 °C. There are also two additional hot water boilers in case IIIA, and the total heat output of the heat and power plant is the same as in case IIA. Case IIIB without hot water boilers constitutes the equivalent of case IIB, the difference being that in view of the higher temperature of the boiler feed water, regenerative heating of the feed water to 150 °C was adopted.

System IV has the largest boilers of those considered, namely two pulverised fuel boilers with outlet parameters of 9·7 MPa, 510 °C and one back-pressure turbine with an outlet heat output of 87·2 MJ/s. In case IVA, two hot water boilers are added so that the total heat output of the heat and power plant amounts to 174 MJ/s in hot water. In case IVB without water boilers, the boiler feed water is heated to 220 °C, which, as the capacity of the steam boilers is the same, means a drop in heat output at the back-pressure turbine outlet to about 67·4 MJ/s.

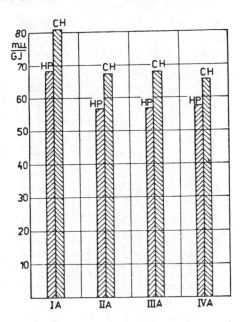

Fig. 4.13   *Comparison of specific costs of heat produced in industrial heat and power and heating plants at T_{ph} = 2000 h/a (case A)*

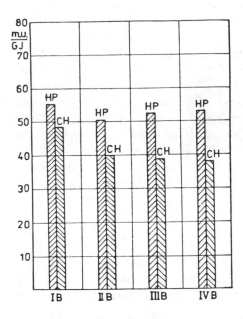

Fig. 4.14   *Comparison of specific costs of heat produced in industrial heat and power and heating plants at T_{ph} = 6000 h/a, without standby boiler (case B)*

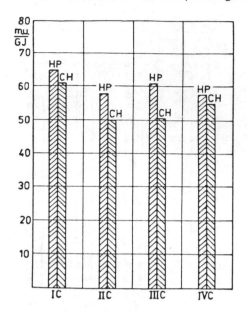

Fig. 4.15   *Comparison of specific costs of heat produced in industrial heat and power and heating plants at T_{ph} = 6000 h/a, with standby boiler (case C)*

Figures 4.13, 4.14 and 4.15 present the results of calculations of heat costs in combined and separate production for the particular model systems, graphically. The specific costs of producing heat in heat and power plants vary between 65 ÷ 81 m.u./GJ in cases A where T_{ph} = 2000 h/a and between 32 ÷ 48 m.u./GJ in case B where T_{ph} = 6000 h/a. In cases C, with the standby boiler, however, the heat costs vary between 49 ÷ 61 m.u./GJ at T_{ph} = 6000 h/a.

The specific costs of producing heat in equivalent heating plants vary between 57 ÷ 68 m.u./GJ in cases A where T_{ph} = 2000 h/a, between 50 ÷ 56 m.u./GJ in cases B where T_{ph} = 6000 h/a and between 58 ÷ 65 m.u./GJ in cases C with the standby boiler. On the other hand, the differences in heat costs vary correspondingly from −13 m.u./GJ to −8 m.u./GJ in cases A, from +7 m.u./GJ to +21 m.u./GJ in cases B and from +3 m.u./GJ to +11 m.u./GJ in cases C. In this case, the profitability condition for combined production constitutes the positive value of the difference in specific heating costs.

Based on the examples mentioned, the following conclusions can be drawn regarding the profitability of combined production of electrical and heat energy in small and medium-sized industrial heat and power plants as compared with separate production of heat energy in equivalent heating plants and electrical energy in equivalent condensing power plants in a power system.

1. With a demand for heat in the form of hot water solely for heating purposes in an industrial plant and a corresponding peak-load utilisation time of about 2000 h/a, industrial heat and power plants with a total heat

output of 43·6 ÷ 174 MJ/s, fitted out with two steam boilers, one back-pressure turbine and two peak-load hot water boilers continue to be unprofitable as compared with equivalent heating plants of the same heat output fitted with hot water boilers only.

2. In view of the above, it can be assumed that in this case the lower limit of profitability of combined heat and power production, where the load is solely for heating, is over a heat output of 175 MJ/s and a corresponding annual production of heat energy of 1260 TJ/a.

3. If there is a demand for steam solely for processing purposes in an industrial plant and the corresponding utilisation time of the peak-load is about 6000 h/a, an industrial heat and power plant with a total heat output of 20·3 MJ/s and an annual heat energy production of 440 TJ/a, in which two steam boilers and one back-pressure turbine are installed, is on the borderline of profitability as compared with an equivalent heating plant with low steam paramaters.

4. In view of the above, it can be assumed that the lower limit of profitability of combined heat and power production, where the load is only for process heat, is around an annual production of heat energy of about 800 TJ/a.

5. If there is a demand for heat for both heating and processing purposes, also a resultant peak-load utilisation time, the lower limit of profitability of combined heat and power production is between the above defined extreme values. In any case, it can be assumed that if the annual demand for heat in an industrial plant is below 800 TJ/a, the inexpediency of combined heat and power production is unquestionable.

6. The relatively substantial rise in the lower limit of profitability of combined production in Poland in recent years was due, primarily, to a distinct increase in the specific investment costs in industrial heat and power plants.

7. The influence of changes in process steam pressure at the outlet of the back-pressure turbine, on the specific costs of producing heat energy in combined heat and power production is relatively slight. The lowering of back-pressure from 0·5 MPa to 0·25 MPa causes the heat costs to drop, an increasing of the back-pressure from 0·5 MPa to 0·8 MPa causes a rise in the heat costs. In view of this, increasing the back-pressure always corresponds to a lowering of the positive difference in heat costs.

8. The influence of the standby boiler on the heating costs in a heat and power plant is distinct. In cases with three boilers instead of two, with the same heat load as in cases with two boilers, the specific heating costs are higher by 25 ÷ 40%. However, as systems with one standby boiler were adopted in equivalent heating plants, the positive differences in heating costs do not, for all practical purposes, depend on whether a third standby boiler is added to the two steam boilers.

Chapter 5

Optimisation problems in combined systems

5.1 Choice of optimum steam outlet parameters in combined heat and power plants

5.1.1 Optimum parameters of steam supplying process heat receivers

When supplying the receivers of process steam from the outlets of back-pressure turbine sets installed in industrial heat and power plants, the problem is to choose optimum steam outlet parameters, i.e. pressure p_b and temperature t_b at the turbine outlet. The outlet pressure (back-pressure) p_b depends on the one hand on the pressure required by the steam receivers and the drop in pressure foreseen in the pipelines supplying steam to the receivers, and on the other has a decisive effect on the electrical power generated by the turbogenerator set at a given heat output delivered from the back-pressure outlet. The outlet temperature t_b, however, depends not only on the pressure p_b, but also on the inlet parameters p_0, t_0 and the internal efficiency of the turbine η_i. These relationships have been presented previously in section 2.1 and in Fig. 2.1. The main problem is the choice of pressure p_b, as it is most frequently the condensing heat of this steam at a constant pressure which is utilised in receivers of process steam.

If the temperature t_b at the turbine outlet were equal to the saturation temperature at pressure p_b, partial condensing of steam would take place in the steam pipelines leading from the turbine to the receivers. In order to avoid this, we assume a temperature t_b higher than the saturation temperature t_s by $\Delta t = 20 \div 30$ K when a temperature close to saturation is obtained at the inlet to the receivers. Should, however, the temperature t_b be much higher, superheated steam would flow into the receivers, which, in turn, would lead to unnecessary increase in the heating surface of these receivers, in view of the lower values of the coefficients of heat absorption from the superheated steam.

Fig. 5.1 presents the relationship between the combined production index σ and back-pressure p_b at different inlet parameters p_0, t_0. This figure illustrates

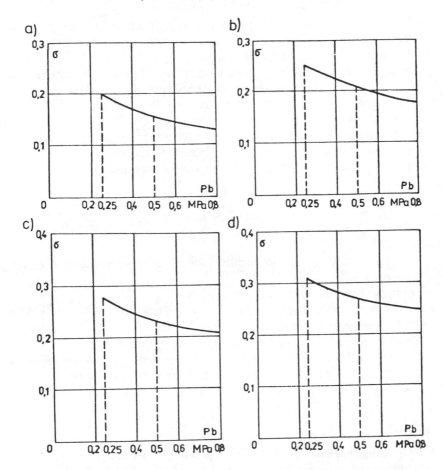

Fig. 5.1 *Combined production indices for back-pressure heat and power plants, depending upon the parameters of inlet and outlet steam: (a) $p_0 = 3\cdot4$ MPa, $t_0 = 435\,°C$; (b) $p_0 = 6\cdot4$ MPa, $t_0 = 465\,°C$; (c) $p_0 = 8\cdot8$ MPa, $t_0 = 500\,°C$; (d) $p_0 = 12\cdot5$ MPa, $t_0 = 535\,°C$*

how substantially the combined production index and thus the electrical power of the back-pressure turbine drops, when the back-pressure p_b increases. Particular attention should, therefore, be paid to the possibility of lowering the back-pressure at the cost of lowering the unnecessary pressure reserve between the pressure at the turbine outlet and the required pressure at the receivers of process steam.

The consumers of this steam frequently demand too high a pressure at the outlet from the heat and power plant, wanting in this way to diminish the heat exchange surface in the steam receivers and increase the safety indices in respect of the required temperature of steam and the drop in pressure between the turbine and the receiver. At a given, required steam pressure at the receiver inlet, the optimum outlet pressure $p_{b\ opt}$ in a heat and power plant, is thus the lowest possible pressure for technical reasons, at which the highest possible electrical energy output is obtained.

In certain cases, it is even possible to carry out a profitable change in the designed receiver, by increasing its heating surface, if this enables a lowering of steam pressure at the back-pressure turbine outlet. For such a change to be profitable, the increase in costs of the receiver should be lower than the increased value of the additionally generated back-pressure electrical energy. Assuming that the increase in heating surface of the receiver results solely in an increase it its fixed costs, this condition can be defined in the form of an inequality

$$(r + r_o)\ \Delta K_i < k_{Eb}\ \Delta\sigma\, H_q \qquad (5.1)$$

when:

r – annual reproduction rate, determined for the assumed accumulation rate p as in section 4.2.2;

r_o – coefficient of annual fixed operating costs;

ΔK_i – increment of receiver investment costs related to the lowering of the required pressure of process steam;

k_{Eb} – unit value of back-pressure electrical energy;

$\Delta\sigma$ – combined production index increment related to the lowering of the pressure of outlet steam in a back-pressure heat and power plant;

H_q – heat energy delivered to the steam receiver.

5.1.2 Optimum parameters of steam supplying district heat exchangers

When supplying district heat exchangers, i.e. the hot water heaters for the district heating from back-pressure or extraction turbines installed in district heating or industrial heat and power plants, the problem arises as to the choice of optimum extraction steam pressure p_e. This pressure depends upon the temperature to which the water is heated in the base-load exchanger or set of base-load exchangers.

Described in Chapter 3 are the graphs of the regulation of heat delivered in the hot water from a heat and power plant. When applying qualitative regulation by means of the district heating water temperature at a constant rate of water flow, two methods of dividing the heat load between the base-load and peak-load exchangers are possible, depending upon the variable external temperature:

– the operation of the base-load exchanger with constant water temperature after the exchanger and variable heat output delivered by this exchanger;
– the operation of the base-load exchanger with variable water temperature after the exchanger and constant heat output delivered by this exchanger.

In the first case, the pressure p_e of the steam supplying the base-load exchanger is constant, as the saturation temperature t_{es} of this steam depends upon the temperature τ_{be} of the district heating water after the base-load exchanger

$$t_{es} = \tau_{be} + \Delta\tau \tag{5.2}$$

where $\Delta\tau$ – temperature difference in the base-load exchanger.

In the second case, the pressure p_e varies between $p_{e\,min}$ and $p_{e\,max}$, corresponding to the change of τ_{be} in the function of the external temperature τ_{ex}. In both cases, however, the problem of choosing the optimum value of pressure p_e leads to the choice of the optimum value of the combined base-load factor α_p, which was mentioned in Chapter 3. At specific temperatures of district heating water at peak periods, a relationship exists between the combined base-load factor at the peak-load time α_p and the temperature of the water behind the base-load exchanger

$$\alpha_p = \frac{\tau_{be} - \tau_{2p}}{\tau_{1p} - \tau_{2p}} \tag{5.3}$$

where:
τ_{1p} – water temperature at the outlet from the heat and power plant at peak period;
τ_{2p} – return water temperature at the inlet to the heat and power plant at peak period;
τ_{be} – water temperature behind the base-load exchanger at peak period.

Example 5.1
Determine the shares of the base-load and peak-load exchangers at peak-load in the heat and power plant, if the district heating water temperatures at peak period amount to $\tau_{1p} = 150\ °C$, $\tau_{2p} = 70\ °C$, and the temperature difference in the base-load exchanger $\Delta\tau = 9\ K$, if the extraction steam pressure p_e varies between $100 \div 250$ kPa.

Solution:
At pressure $p_e = 100$ kPa the saturation temperature $t_{es} = 99\ °C$, thus

$$\tau_{be} = 99 - 9 = 90\ °C$$

and the share of the base-load exchanger in covering the peak-load is

$$\alpha_p = \frac{90 - 70}{150 - 70} = 0\cdot25$$

and the share of the peak-load exchanger (or peak-load boiler) is

$$1 - \alpha_p = 0\cdot75$$

At pressure $p_e = 250$ kPa the temperature $t_{es} = 127\,°\text{C}$, thus

$$\tau_{be} = 127 - 9 = 118\,°\text{C}$$

$$\alpha_p = \frac{118 - 70}{150 - 70} = 0{\cdot}60$$

$$1 - \alpha_p = 0{\cdot}40$$

It results from this example that at a given temperature of the district heating water at peak period, the optimum outlet (extraction) steam pressure in a heat and power plant $p_{e\ opt}$ is that at which the combined base-load factor α_p, defined as the part played by the base-load exchanger in the peak-load, attains the optimum value. With the temperature $\tau_{1p} = 150\,°\text{C}$, $\tau_{2p} = 70\,°\text{C}$ assumed in Poland and a combined base-load factor $\alpha_{p\ opt} \approx 0{\cdot}4$, the optimum outlet (extraction) steam pressure amounts to $p_{e\ opt} \approx 130$ kPa.

As, however, the operation of the base-load exchanger with variable water temperature behind the exchanger constitutes a more advantageous solution in view of the higher attainable production of back-pressure electrical energy, the outlet pressure during the heating season changes to $p_{e\ min} \approx 40$ kPa. The mean outlet pressure during the season, amounts, in this case, to $p_e \approx 55$ kPa. If, however, the heat and power plant does not cover the demand for hot water for utility purposes, the outlet pressure can be further lowered to $p_{e\ min} \approx 20$ kPa. This corresponds to the three-stage heating of the district heating water, the diagram of which was presented in Fig. 3.12.

5.2 Choice of optimum steam inlet parameters in combined heat and power plants

5.2.1 Technical limits of steam inlet parameters
Section 2.1 gave the relationships between the inlet parameters, i.e. pressure and temperature of steam at the back-pressure turbine inlet, and the outlet parameters at the turbine outlet. These relationships, given different inlet parameters and internal efficiencies of a back-pressure turbine, were illustrated by the graphs in Fig. 2.1.

As can be seen there, the outlet temperature t_b drops distinctly with the improvement in internal efficiency η_i at the given inlet parameters p_0, t_0 and back-pressure p_b. It is also known that the steam temperature t_b at the back-pressure turbine outlet should be higher than the saturation temperature t_s at a given pressure by at least $20 \div 30$ K to avoid condensing of steam in the steam pipelines.

The maximum inlet pressure p_0 in a back-pressure cycle, which at the given inlet temperature t_0 and given internal efficiency η_i can be at least such that the corresponding outlet temperature amounted to $t_b = t_s$ or $t_b = t_s + \Delta t$, can be determined on this basis. It has been found that the maximum pressure p_0 so determined, is usually higher, at the given inlet temperature t_0 and given internal efficiency η_i, than the maximum inlet pressure in a condensing cycle at the same inlet temperature.

In order to determine the maximum parameters of inlet steam in back-pressure heat and power plants numerically, a series of heat cycle calculations [38] were carried out. The following data were assumed:

three values of inlet steam temperature t_0

$$435;\ 500;\ 565\ ^\circ\text{C}$$

four back-pressure values p_b

$$0{\cdot}25;\ 0{\cdot}5;\ 0{\cdot}8;\ 1{\cdot}0\ \text{MPa}$$

two outlet steam temperature levels

$$t_b = t_s \text{ and } t_b = t_s + 20\ \text{K}$$

internal efficiency of back-pressure turbine $\eta_i = 0{\cdot}8$.

The results, in the form of relationships $p_o = f(p_b)$ at different t_0 and t_b are presented in Fig. 5.2. Inserted in this figure are also horizontal lines

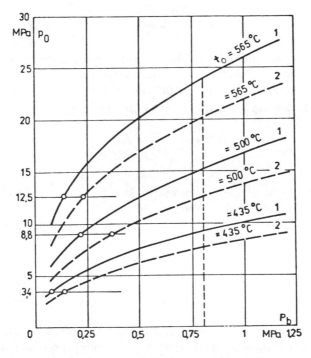

Fig. 5.2 *Maximum inlet pressures in a back-pressure heat and power plant, depending on the outlet parameters when $\eta_i = 0{\cdot}8$*
1 – at an outlet temperature t_b equal to the saturation temperature, 2 – at an outlet temperature t_b by 20 K higher than the saturation temperature

corresponding to the most frequently assumed, standardised values of inlet pressure

$$p_0 = 3.4 \text{ MPa at } t_0 = 435 \text{ °C}$$
$$p_0 = 8.8 \text{ MPa at } t_0 = 500 \text{ °C}$$
$$p_0 = 12.5 \text{ MPa at } t_0 = 565 \text{ °C}$$

As can be seen from the graphs, at $t_b = t_s + 20$ K, the admissible maximum inlet pressures are usually higher than the typical pressures given above, in the whole back-pressure range, from $p_b = 0.37$ MPa. For a back-pressure of $p_b = 0.25$ MPa, however, this condition is fulfilled at $t_0 = 435$ °C and $t_0 = 565$ °C, but not at $t_0 = 500$ °C. Hence, at a pressure of $p_0 = 8.8$ MPa, it is worth while increasing the temperature to $t_0 = 535$ °C.

When choosing the steam inlet parameters in a back-pressure heat and power plant, it can be required that the production of electrical energy be as high as possible at the given back-pressure. This leads to the choice of parameters on the criterion of maximisation of energy generated in a back-pressure cycle, which can be formulated in two ways:

(1) as the criterion of maximum net value of combined production index:

$$\sigma_n = \frac{P_{bn}}{Q_{bn}} = max \tag{5.4}$$

where:
σ_n – the net value of combined production index in a back-pressure system;
P_{bn} – net electrical power generated in a back-pressure cycle;
Q_{bn} – net heat output delivered from the back-pressure outlet;

(2) as the criterion of maximum net index of electrical energy generation

$$\varrho = \frac{P_{bn}}{D_b} = max \tag{5.5}$$

where:
ϱ – index of electrical energy generation in a back-pressure cycle;
D_b – steam flow of a back-pressure turbine.

In the first case the formula (5.4), concerning the combined production index, can be transformed into

$$\sigma_n = \frac{P_{bn}}{Q_{bn}} = \frac{\Delta h_b \, \eta_{em} \, (1 - \varepsilon)}{\Delta h_q} \tag{5.6}$$

here:
ε – share of auxiliary power demand for the heat and power plant;
remaining notations – as in section 1.2 in connection with the formulae (1.1–1.5).

In order that the auxiliary power demand ε be dependent on the steam inlet parameters in the heat and power plant, it was assumed that the feed-water pump has a deciding share. The power consumption of the pump motor taking into account its efficiency η_{me} and the efficiency of the auxiliary transformer η_{tre} was determined, it then being assumed that the remaining consumption for auxiliaries is proportional to the power consumed by the feed-water pump motor.

To determine the relationship $\sigma_n = f(p_0)$ at various inlet temperatures t_0 and back-pressures p_b, calculations of several back-pressure cycles were carried out, taking into account the following assumptions:

internal efficiency of back-pressure turbine $\hspace{2cm} \eta_i = 0{\cdot}8;$
electromechanical efficiency of turbogenerator set $\;\; \eta_{em} = \eta_m\,\eta_t\,\eta_g = 0{\cdot}95;$
feed-water pump efficiency $\hspace{4cm} \eta_{fw} = 0{\cdot}75;$
efficiency of electrical auxiliary equipment $\hspace{1.5cm} \eta_{au} = \eta_{tre}\,\eta_{me} = 0{\cdot}9$

The calculations were carried out assuming the same parameters as previously, i.e.

three temperatures of inlet steam t_0: 435°; 500°; 565 °C
three values of back-pressure p_b: 0·25; 0·5; 0·8 MPa.

Fig. 5.3 presents the results for $t_0 = 435\ °C$. As can be seen from the graphs, the inlet pressure corresponding to $\sigma_n = \text{max}$, is very high and amounts to about 23 MPa in the case in question. This pressure increases slightly with increased back-pressure. At the same time, it is, as a rule, substantially higher than the maximum pressure admissible in view of the temperature of the inlet steam, defined previously in Fig. 5.2.

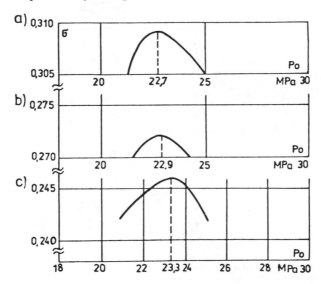

Fig. 5.3 *Dependence of the net combined production index for a back-pressure heat and power plant on the steam inlet parameters and back-pressure when $t_0 = 435\ °C$: (a) $p_b = 0{\cdot}25$ MPa; (b) $p_b = 0{\cdot}5$ MPa; (c) $p_b = 0{\cdot}8$ MPa*

In the second case, the formula (5.5) referring to the index of electrical energy generation in a back-pressure cycle, can be presented in the form

$$\varrho = \frac{P_{bn}}{D_b} = \Delta h_b \, \eta_{em} \, (1 - \varepsilon) \tag{5.7}$$

here the notations are as in the formula (5.6).

Calculations of relationships $\varrho = f(p_0)$ at various inlet temperatures t_0 and back-pressures p_b were carried out assuming the same efficiencies η_i, η_{em}, η_{fw}, η_{au} as previously.

The results are presented in Fig. 5.4 in the form of graphs $p_0 = f(p_b)$ at various temperatures t_0, where the dependencies in relation to inlet pressure, corresponding to the condition $\varrho = \max$, are presented by continuous lines, and those concerning maximum admissible pressure due to the temperature of the outlet steam – by broken lines. As can be seen from the graphs, the inlet pressure defined on the basis of this criterion of maximisation of generated energy, is also higher than the admissible for technical reasons.

The criteria of choice of inlet steam parameters in heat and power plants discussed, are based solely on technical factors (maximisation of production) not taking into account the economic aspects of the choice. Therefore they cannot be accepted as the basis for optimisation of back-pressure cycles.

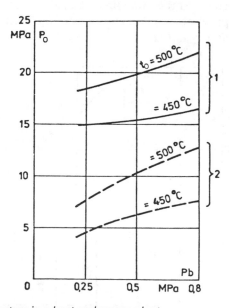

Fig. 5.4   *Inlet parameters in a heat and power plant*
1 – optimum on account of the power output obtained from a unit of back-pressure steam flow,
2 – maximum on account of the temperature of outlet steam from the back-pressure turbine

5.2.2 Choice based on efficiency of heat and power plants

The next criterion on which the choice of optimum parameters of steam can be based, is the overall efficiency of heat and power plants. In a condensing power plant, the net overall efficiency, or the net consumption of fuel per unit of electrical energy generated, which is strictly related to this, are considered to constitute the basic indices which afford the most synthetic characterisation of the technical standard of energy generation attained. In a district heating plant, the analogical index is the net consumption of fuel per unit of heat energy generated.

Such indices are, of course, necessary when assessing the design of a power or a district heating plant and later to estimate the level of operation, but do not, however, suffice as a criterion. As is known, the achievement of greater efficiency, and thus lower operational costs, is usually related to higher investment outlays, which result in higher annual fixed costs. The final economic criteria required to assess the effectiveness of generating energy, thus constitute the total net specific energy costs, which consist of both fixed costs, depending upon the amount of investment outlays and variable operational costs, depending on overall efficiency.

The determining of the overall efficiency of the combined generation of electrical and heat energy in a heat and power plant is one of the most difficult theoretical problems in power engineering. Hence several formulations presented in literature give rise to various doubts. For some time now, attempts have been made in the proposed formulae for the overall efficiency of a heat and power plant, to take into account in one index, the complex character of combined energy generation and include both kinds of energy. On the other hand, however, when introducing the reverse concept, which is the specific fuel consumption, there has always been a division into the fuel consumption falling to the electrical energy and that falling to the heat energy, as the idea of a general specific consumption would be too artificial.

The formula (1.43) for net overall efficiency of a heat and power plant given previously in section 1.3, is based on the physical principle of energy balance resulting from the *first law of thermodynamics* and leads to the conversion of net electrical power P_n into an equivalent heat output, which is added to the net heat output Q_n delivered from the heat and power plant.

Reservations in respect of the formula (1.43) consist mainly of the fact that in the case of diminishing the electrical power of a heat and power plant with simultaneous increase in heat output, the overall efficiency defined by this formula increases as the result of lower mechanical and electrical losses in the turbogenerator set. In an extreme case, where $P_n = 0$, this efficiency attains maximum value, equal to the partial efficiency of net heat energy generation determined by the formula (1.45).

In view of the difficulties mentioned, in determining the overall efficiency of combined generation of energy, based on the *first law of thermodynamics*, there arose the determination of partial efficiencies correspondingly related to the generation of both types of energy, which were discussed in detail in section 1.3.1.

N. Gašparović [13] submitted an interesting proposal for the modification of the existing formula for the overall efficiency of a heat and power plant, simultaneously putting forward the known complaints against the calculation of efficiency according to the formula (1.43). In his opinion, the proper determination of the efficiency of combined generation should fulfil the following conditions:

1. The efficiency of a cycle without losses should be equal to unity

$$(\eta)_{\Delta Q=0} = 1 \qquad (5.8)$$

2. The efficiency of a cycle in which losses occur, should be less then unity

$$(\eta)_{\Delta Q} < 1 \qquad (5.9)$$

3. If the sum of the net electrical and heat output generated is constant

$$P_n + Q_n = C \qquad (5.10)$$

then the increase of electrical power should correspond to the increase in efficiency and the increase of heat output to the corresponding lowering of efficiency

$$\left(\frac{\partial \eta}{\partial P_n}\right)_C > 0 \qquad (5.11)$$

$$\left(\frac{\partial \eta}{\partial Q_n}\right)_C < 0 \qquad (5.12)$$

4. The required formula for the efficiency of a heat and power plant should also be good for extreme cases, i.e. when generating electrical energy only, in condensing power plants (CP) and heat energy only, in district heating plants (HP).

The formula for the heat and power plant efficiency drawn up on the basis of the four cases presented according to Gašparović has the form of

$$\eta_G^{CH} = \frac{P_n + Q_n \dfrac{P_n + Q_n}{Q_0}}{Q_0} \qquad (5.13)$$

For extreme cases the corresponding expression is:

for a condensing power plant ($Q_n = 0$)

$$\eta_G^{CP} = \eta_n^{CP} = \frac{P_n}{Q_0} \qquad (5.14)$$

thus in accordance with the practice hitherto;

for a district heating plant ($P_n = 0$)

$$\eta_G^{HP} = \left(\frac{Q_n}{Q_0}\right)^2 \qquad (5.15)$$

This formula differs from the existing, but affords a certain preference to installations and systems with a greater Q_n/Q_0 ratio, that is to say, heating plants with a higher efficiency in the existing sense.

It can be shown that a relationship exists between the overall efficiency η_G^{CH} of a heat and power plant according to Gašparović from the formula (5.13) and the efficiency of generating electrical energy η_n^{CP} according to the formula (5.14)

$$\eta_G^{CH} = \eta_n^{CP} \left[1 + \frac{1}{\sigma_n}\left(1 + \frac{1}{\sigma_n}\right)\eta_n^{CP}\right] \qquad (5.16)$$

whereas the relationship

$$\eta_n^{CH} = \eta_n^{CP}\left(1 + \frac{1}{\sigma_n}\right) \qquad (5.17)$$

resulted from the existing definition of efficiency η_n^{CH} according to the formula (1.43).

Further modification of the notion of efficiency is connected with the introduction of exergy to the analysis and assessment of thermodynamic processes. It was mainly W. Fratzscher [12] who was engaged in this problem and in Poland – S. Ochęduszko [63], J. Pientka [66] and J. Szargut [78, 80].

The exergetic efficiency of the heat cycle in a back-pressure heat and power plant can be expressed by means of the following relationship

$$\eta_{ex}^{CH} = \frac{e_{Eb} + e_{Hb}}{e_{sc}} \qquad (5.18)$$

where:
e_{Eb} – exergy of electrical energy generated in a back-pressure cycle;
e_{Hb} – exergy of heat delivered from a heat and power plant in back-pressure steam;
e_{sc} – exergy of heat supplied to the steam cycle in the heat and power plant boilers.

Here, in accordance with the previous denotations

$$e_{Eb} = \frac{P_{bn}}{D_b} \qquad (5.19)$$

$$e_{Hb} = \frac{Q_{bn}}{\varepsilon_0 D_b} \qquad (5.20)$$

$$e_{sc} = \frac{Q_0 \eta_{sb}}{\varepsilon_0 D_b} \qquad (5.21)$$

whereas ε_0, ε_q denote correspondingly the exergetic value of heat supplied and delivered, expressed by means of the absolute temperature

$$\varepsilon_0 = \frac{T_{av0}}{T_{av0} - T_{amb}} \tag{5.22}$$

$$\varepsilon_q = \frac{T_{avq}}{T_{avq} - T_{amb}} \tag{5.23}$$

where:

T_{av0} – the mean absolute temperature at which is supplied to the heat carrier, i.e. water and steam in heat and power plant boilers;

T_{avq} – mean absolute temperature at which heat is delivered by outlet steam from heat and power plant;

T_{amb} – assumed ambient absolute temperature;

η_{sb} – efficiency of steam boilers in the heat and power plant.

Taking into consideration the various notions of the efficiency of electrical and heat energy in a heat and power plant as presented, an attempt was made to choose the optimum inlet steam parameters basing on the criterion of maximisation of efficiency. Taking into account the same assumptions as previously, calculations were carried out for $t_0 = 500$ °C and $p_b = 0.5$ MPa, and for the range of inlet pressure variability $15 \leqslant p_0 \leqslant 25$ MPa.

It was found that the efficiencies η_n^{CH}, η_G^{CH}, and η_{ex}^{CH} change very little if inlet parameters change, and the maximum efficiency occurs at $p_0 \approx 20$ MPa. The optimum inlet pressure is thus approximately that obtained previously in section 5.2.1 based on the criterion $\varrho = $ max, at the same back-pressure $p_b = 0.5$ MPa.

As previously it can thus be stated that the inlet pressure chosen on the basis of maximisation of the efficiency of a back-pressure heat and power plant is higher than the pressure permitted for technical reasons, i.e. in view of the outlet temperatue t_b, and cannot thus be considered to be the optimum pressure.

5.2.3 Choice based on annual costs of heat and power generation

As opposed to the criteria of choosing heat and power plant parameters described in sections 5.2.1 and 5.2.2, which took into account only technical limits resulting from the thermodynamics of heat cycles, or resulted from different formulations of the notion of efficiency, the principle of minimising total, annual costs of generating electrical and heat energy in a combined system [38] has been assumed in this section as the basic economic criterion enabling the optimum choice of inlet parameters in heat and power plants, in the following considerations.

The total annual costs K in a heat and power plant are the function of many variables which primarily include inlet and outlet parameters, power output of installations, specific investment costs and accumulation rate, on which the reproduction rate depends. If the inlet temperature t_0 becomes dependent on the pressure p_0, then instead of two variables p_0, t_0, only one variable p_0 can be considered. The minimisation of annual costs can then be written as

$$\frac{\partial K}{\partial p_0} = 0 \tag{5.24}$$

where, after finding the value $p_{0\,opt}$, fulfilling the equation (5.24), it should, for accuracy, be stated whether the extremum found is really the minimum of function $K(p_0)$.

As the definition of the function $K(p_0)$ in analytical form is, for all practical purposes, impossible, a method was applied which utilised the difference in annual costs ΔK between combined production and an equivalent separate production according to the formula (4.56).

If the value p_{01} of the variable p_0 studied corresponds to the annual costs of the heat and power plant K_1, and the increment of the variable p_0, amounting to dp_0, corresponds to the increment in cost dK, then the annual cost of the heat and power plant corresponding to an inlet pressure $p_{01} + dp_0$ will amount to $K_1 + dK$. However, the increase of inlet pressure by dp_0 in the heat and power plant is accompanied by a slight change in heat and energy output supplied to the heat consumers and an essential change in the electrical power and energy produced in a combined system, and thus a change in the production effect of the heat and power plant.

In order to retain the conditions of comparison of the variant before the change with that after the change, the equivalent of the effect, calculated in the form of increments of equivalent costs of a condensing power plant dK_{CP} and the equivalent heating plant dK_{HP}, should be subtracted from the annual cost. Thus the resulting increment in the annual cost of a heat and power plant will amount to

$$K_1 + dK - dK_{CP} - dK_{HP} - K_1 = -d(\Delta K) \tag{5.25}$$

as it results from the definition of ΔK given in section 4.2.2 that

$$d(\Delta K) = d(K_{CP} + K_{HP} - K) = dK_{CP} + dK_{HP} - dK \tag{5.26}$$

It is thus illustrated that instead of studying the minimum annual costs of a heat and power plant, it is sufficient to define the maximum difference of costs ΔK, or – which amounts to the same – the maximum value of the ratio $\Delta K/K_{HP}$, to find the value $p_{0\,opt}$, as it can be assumed with fairly good approximation, that the heat output supplied from a back-pressure heat and power plant at a given back-pressure p_b, for all practical purposes does not depend on the changes in inlet pressure p_0 and thus the annual costs of an equivalent power plant K_{HP} do not depend on this either. What will change with the change in p_0 will be the annual costs of the heat and power plant K_{CH} and those of an equivalent condensing power plant K_{CP}.

Examples of calculations of optimum parameters of inlet steam in back-pressure heat and power plants, based on the above criterion

$$\frac{\Delta K}{K_{HP}} = \max \qquad (5.27)$$

were carried out on a computer. In view of this, a programme was designed which enables the values of the criterial function for six inlet steam parameters and three back-pressure values, depending upon one variable value, e.g. the peak-load utilisation time T_{tp}, assuming that the remaining variables are fixed, to be stored in the computer memory. Next, for a given column, which corresponds to the required value of the variable studied, e.g. required utilisation time T_{tp}, it is possible to find the maximum value of $\Delta K/K_{HP}$ and determine the line in which this maximum value occurs, in the matrix stored. After carrying out calculations for all the back-pressure values, the computer prints out the optimum steam parameters and the maximum values of $\Delta K/K_{HP}$ corresponding to these.

Applying the programme described, calculations were carried out for a heat and power plant with a mixed heating and process load, having identical boilers with outputs of 16, 32, 70 and 140 t/h and back-pressure turbines of corresponding output. Examples of calculation results for heat and power plants with two boilers of 32 and 70 t/h, with a share of process load in the peak output $u_{tp} = 0.5$ and back-pressure $p_b = 0.25$ and 0.8 MPa are shown in Fig. 5.5. In this figure the intermediate values corresponding to $p_b = 0.5$ MPa have been omitted. The utilisation time T_{tp} has been varied between 2000 h/a and 6000 h/a.

For these examples of calculations, the parameters assumed were those of live steam at the boiler outlet and inlet to the back-pressure turbine, as in Table 2.1. Each inlet pressure value has been given two different inlet temperatures, thus increasing the total number of variants.

As a result, a group of curves was obtained, these representing the relationships

$$\frac{\Delta K}{K_{HP}} = f(p_0) \qquad (5.28)$$

at a temperature t_0 correspondingly correlated with the pressure p_0. The utilisation time of the peak demand for process heat T_{tp} is one of the parameters in the group of curves in Fig. 5.5.

Although the resultant curves are fairly flat, the maximum values of the functions studied and the corresponding optimum parameters, which are denoted with circles in Fig. 5.5, can be determined in all cases. The functions corresponding to the changed inlet temperature values t_0' are shown with dotted lines.

As it results from the calculations in this example, the optimum inlet parameters amount to from 6.4 MPa, 465 °C to 8.8 MPa, 535 °C – depending upon the utilisation time, back-pressure and heat output in the heat and power plant. In the case of increased back-pressure, the positive values of the difference in costs ΔK decrease, as the electrical back-pressure power produced on a corresponding steam flow decreases and the benefits from combined production diminish. Simultaneously, the optimum steam parameters fall. On the other hand, if the utilisation time of the heat load increases, then savings resulting from combined production rise and the optimum parameters increase correspondingly.

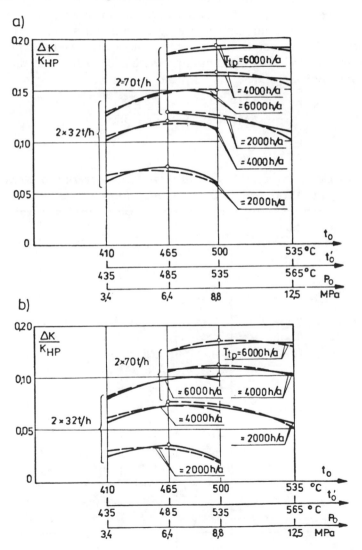

Fig. 5.5 *Optimum inlet parameters in a back-pressure heat and power plant depending upon the back-pressure and the utilisation time of the peak demand for process heat (a) $p_b = 0.25$ MPa, $u_{tp} = 0.5$; (b) $p_b = 0.8$ MPa, $u_{tp} = 0.5$*

5.2.4 Factors influencing optimum steam inlet parameters in back-pressure heat and power plants

Resulting from the above calculations [38], the following factors have a decisive influence on the value of the optimum inlet pressure:

1. *Size of boiler and turbine machinery.* The optimum parameters depending upon the output of the boiler units for variants with two boilers per back-pressure turbine are listed in Tables 5.1 and 5.2. Higher parameters are usually suitable for larger units.
2. *Utilisation time of peak heat demand.* As can be seen from Table 5.1, increased utilisation time causes the optimum steam parameters to increase. For example heat and power plant with two boilers of 16 t/h at an index of $u_{tp} = 0.5$ and back-pressure $p_b = 0.25$ MPa, the optimum parameters are:

$$3.4 \text{ MPa}, 435 \,^{\circ}\text{C at } T_{tp} = 3000 \text{ h/a}$$
$$6.4 \text{ MPa}, 465 \,^{\circ}\text{C at } T_{tp} = 5000 \text{ h/a}$$

Table 5.1   *Dependence of optimum steam inlet parameters on the utilisation time of process heat demand*

Share of process heat	u_{tp}	–	0·5	0·5	0·5
Back-pressure	p_b	MPa	0·25	0·25	0·25
Rated capacity of steam boiler	D_{sbr}	t/h	16	32	70
No. of boilers	N_{sb}	–	2	2	2
Utilisation time T_{tp}			Optimum steam inlet parameters		
2000 h/a	p_0	MPa	3·4	6·4	6·4
	t_0	°C	435	465	465
3000 h/a	p_0	MPa	3·4	6·4	6·4
	t_0	°C	435	465	465
4000 h/a	p_0	MPa	6·4	6·4	8·8
	t_0	°C	465	465	535
5000 h/a	p_0	MPa	6·4	6·4	8·8
	t_0	°C	465	465	535
6000 h/a	p_0	MPa	6·4	6·4	8·8
	t_0	°C	465	465	535

Table 5.2 *Dependence of optimum steam inlet parameters on steam back-pressure*

Utilisation time	T_{tp}	h/a	5000	5000	5000
Rated capacity of steam boiler	D_{sbr}	t/h	16	32	70
No. of boilers	N_{sb}	–	2	2	2
Share of process heat	u_{tp}	–	0·5	0·5	0·5
Back-pressure p_b		Optimum steam inlet parameters			
0·25 MPa	p_0	MPa	6·4	6·4	8·8
	t_0	°C	465	465	535
0·5 MPa	p_0	MPa	6·4	6·4	8·8
	t_0	°C	465	465	535
0·8 MPa	p_0	MPa	6·4	6·4	8·8
	t_0	°C	465	465	535
Share of process heat	u_{tp}	–	0·8	0·8	0·8
Back-pressure p_b		Optimum steam inlet parameters			
0·25 MPa	p_0	MPa	6·4	8·8	8·8
	t_0	°C	465	535	535
0·5 MPa	p_0	MPa	6·4	8·8	8·8
	t_0	°C	465	535	535
0·8 MPa	p_0	MPa	6·4	8·8	8·8
	t_0	°C	465	535	535

3. *Pressure of outlet steam from a back-pressure turbine.* Increased back-pressure results in lower profitability of combined production and the lowering of optimum inlet parameters. For example, for a heat and power plant as in Section 2 and at a utilisation time $T_{tp} = 4000$ h/a it is

6·4 MPa, 465 °C at $p_b = 0·25$ MPa
3·4 MPa, 435 °C at $p_b = 0·5$ and 0·8 MPa

4. *Ratio of process/heating heat demand.* An increase in the share of demand for process heat u_{tp} in the total peak demand for heat causes an increase in the optimum steam parameters, on condition that the utilisation time T_{tp} is sufficiently long, as then, the increase of u_{tp} causes an increase in the resultant utilisation time. For example, in the case of a heat and power plant as in Section 2, where $T_{tp} = 4000$ h/a and $p_b = 0.5$ and 0.8 MPa we have

$$3.4 \text{ MPa}, 435\,°C \text{ when } u_{tp} = 0.5$$
$$6.4 \text{ MPa}, 465\,°C \text{ when } u_{tp} = 0.8$$

5. *Price of fuel.* An increase in the price of equivalent coal in a heat and power plant c_f also results in a corresponding increase in the optimum parameters of steam. For example, in the case of a heat and power plant as in Section 2 it will be

$$3.4 \text{ MPa}, 435\,°C \text{ if } c_f < 340 \text{ m.u./t*}$$
$$6.4 \text{ MPa}, 465\,°C \text{ if } c_f > 340 \text{ m.u./t}$$

6. *Accumulation rate.* An increase in the accumulation rate p, on which the reproduction rate depends, causes a lowering of optimum parameters of steam. For example, in the case of a heat and power plant as above, this would be

$$6.4 \text{ MPa}, 465\,°C \text{ if } p < 0.16$$
$$3.4 \text{ MPa}, 435\,°C \text{ if } p > 0.16$$

An important economic value on which the limits of profitability of combined generation of energy as compared with separate generation and the optimum parameters of steam depend, is the ratio of specific fixed costs in the heat and power plant to the price of heat contained in the fuel. This ratio is called the fuel equivalent of fixed costs and is defined by the formula

$$R = \frac{k_n r_c}{c_f T_{rh}} \tag{5.29}$$

in which:
k_n – specific investment cost in the heat and power plant, related to the rated heat output;
r_c – coefficient defining the ratio of annual fixed costs to the investment cost;
c_f – price of heat contained in the fuel;
T_{rh} – utilisation time of rated heat output.

An increase in the fuel equivalent R, which is an undimensional relative value, causes a drop in the profitability of the heat and power plant and a lowering of the optimum parameters.

---

* m.u. = monetary unit.

This increase can be due to:
- an increase in the specific investment cost k_n;
- an increase in the coefficient of fixed costs r_c;
- a drop in the price of heat contained in the fuel c_f;
- a drop in the utilisation time of heat output T_{rh}.

5.3 Factors determining profitability limits of combined heat and power generation

The difference in annual costs ΔK defined by the formula (4.56) is taken as the basis for analysing the factors which determine the profitability limits of combined production. As the result of annual cost calculations we obtain the criterion in the form of the relative difference of costs between separate and combined production $\Delta K/K^{HP}$, where

$$\frac{\Delta K}{K^{HP}} = \frac{K^{CP} + K^{HP} - K^{CH}}{K^{HP}}$$

As can be seen from the above formula, to determine the relative difference in costs, the absolute difference ΔK is related to the annual costs of heat production in a heating plant. As the denominator in the formula (5.30) is of positive value, to determine the sign of the function $\Delta K/K^{HP}$, it is sufficient to check the sign of the numerator.

If combined production is profitable, then $\Delta K > 0$ and the criterial function has a positive sign and in the opposite case of $\Delta K < 0$ a negative sign is obtained. The zero values of the criterial function correspond to the limiting conditions of profitability of the heat and power plant.

Calculation of the relative difference of costs $\Delta K/K^{HP}$, resulting from the comparison of combined and separate production, was carried out on a computer, based on a special programme to determine the profitability of heat and power plants, which enables fast, multiple repetition of calculations with different variants of demand for heat and electrical energy and different power equipment [38]. This programme enables the carrying out of calculations taking into account the following sets of the main values:

- six pairs of live steam parameters, i.e. pressures and temperatures at the turbine inlet;
- three back-pressure values;
- twelve different combinations of rated capacities of steam boilers in heat and power plants and of the share of process heat in the peak-load heat output.

For each set of values mentioned, the programme enables the checking of the criterial function $\Delta K/K^{HP}$ depending upon one of the following variable values:

- utilisation time of peak-load heat demand for industrial processes T_{tp};
- specific investments c_b for steam boilers installed in a heat and power

plant, corresponding to the specific investments c_t for turbogenerator sets;
- price of equivalent fuel c_f in a heat and power plant and in an equivalent condensing power plant;
- accumulation rate p, the level of which effects the relative value of fixed production costs.

In addition, it is possible to calculate in each variant, such a value of the variable investigated, at which the citerial function assumes zero value, which corresponds to the profitability limit of combined production.

Applying the programme mentioned, calculations have been carried out for back-pressure heat and power plants embracing two identical boilers with rated capacities of 16, 32 and 70 t/h and one back-pressure turbine of corresponding power output. The results of calculations are presented in Figs. 5.6 to 5.10.

Fig. 5.6 presents the dependence of the criterial function on utilisation time $\Delta K/K^{HP} = f(T_{tp})$ for a $2 \cdot 16$ t/h system at an index of $u_{tp} = 0.5$ and three different sets of live steam parameters p_0, t_0, and for the back-pressure p_b correspondingly equal to 0.25 MPa and 0.8 MPa. Utilisation time T_{tp} was taken from 2000 to 6000 h/a with the remaining values maintained at one level.

The limiting values $(T_{tp})_0$ for the particular steam inlet parameters and the points of intersection of curves corresponding to the various parameters are marked with circles. The following conclusions can be reached from such diagrams:

(1) the profitability of combined production increases with the rise in utilisation time T_{tp};
(2) the limit value of utilisation time $(T_{tp})_0$, at which combined production is profitable, increases with the rise in inlet and outlet parameters of the back-pressure turbine;
(3) the optimum values of steam parameters at the turbine inlet in the system of two boilers of 16 t/h at $u_{tp} = 0.5$ amount to 3.4 MPa, 435 °C at a low utilisation time T_{tp}, rising to 6.4 MPa, 465 °C at higher T_{tp} values.

The dependence of the criterial function on the specific investments $\Delta K/K^{HP} = f(c_b, c_t)$ for the system studied, is presented in Fig. 5.7. It was assumed that the lowest value of c_b corresponds to the lowest value of c_t and vice versa, and in the intermediate points the increments of c_t correspond proportionally to the increments of c_b. The remaining variable values remain as in the previous case.

It results from the diagrams that the profitability of combined production drops distinctly with the increase of specific investments in the heat and power plant. The limiting values of specific investment costs $(c_b)_0$ and $(c_t)_0$ above which combined production becomes unprofitable, can be seen from the graphs.

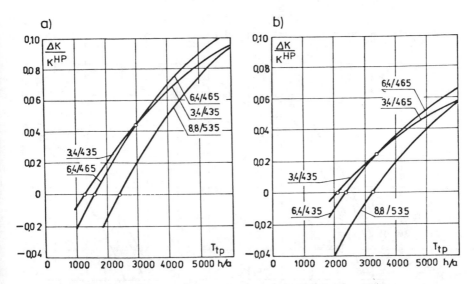

Fig. 5.6   *Profitability of a back-pressure heat and power plant with two boilers of 16 t/h, at $u_{tp} = 0.5$, depending upon the utilisation time of peak-load demand for process heat, when: (a) $p_b = 0.25$ MPa, (b) $p_b = 0.8$ MPa*

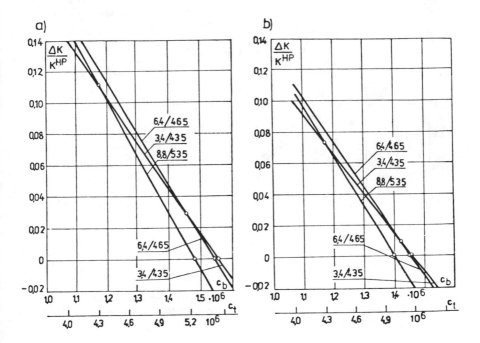

Fig. 5.7   *Profitability of a back-pressure heat and power plant with two boilers of 16 t/h, at $u_{tp} = 0.5$, depending upon the indices of investment costs for the boilers and turbine sets, when: (a) $p_b = 0.25$ MPa, (b) $p_b = 0.8$ MPa*

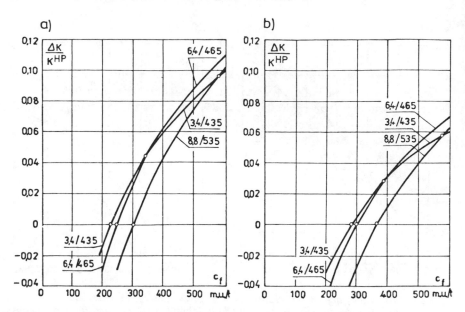

Fig. 5.8 *Profitability of a back-pressure heat and power plant with two boilers of 16 t/h, at $u_{tp} = 0.5$, depending upon the price of fuel, when: (a) $p_b = 0.25$ MPa, (b) $p_b = 0.8$ MPa*

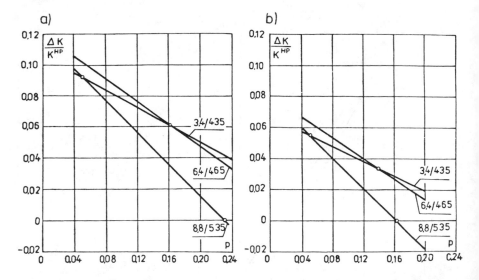

Fig. 5.9 *Profitability of a back-pressure heat and power plant with two boilers of 16 t/h, at $u_{tp} = 0.5$, depending upon the rate of accumulation, when: (a) $p_b = 0.25$ MPa, (b) $p_b = 0.8$ MPa*

The third factor which is decisive as regards the criterial function is the price of fuel. Fig. 5.8 shows the relationship $\Delta K/K^{HP} = f(c_f)$ obtained. It can be seen from the diagrams that combined production which leads, in effect, to saving of fuel, is more profitable the higher the price of fuel. The diagrams show the calculated limiting prices of fuel $(c_f)_0$, which differ with the various parameters of steam. Below these prices, combined production would be unprofitable.

The fourth variable value on which the criterial function depends is the rate of accumulation p. The relationship $\Delta K/K^{HP} = f(p)$ for the heat and power plant studied, is presented in Fig. 5.9. As can be seen from the diagrams, the profitability of combined production depends essentially on the rate of accumulation, i.e. it diminishes with the increase in rate. If the values of p are several times in excess of the value $p = 0.08$ assumed in the remaining terms of calculation, then the savings in variable production costs resulting from lower fuel consumption in combined production are eliminated by the substantial increase in fixed costs which depend on the rate of p.

Similar calculations were carried out for systems of boilers with different rated capacities and back-pressure turbines with a corresponding power output, i.e. for systems of $2 \cdot 32$ t/h and $2 \cdot 70$ t/h. The increase in power output corresponding to the increase in demand for heat and electrical energy, results in the greater profitability of combined production.

Fig. 5.10 presents the cumulative results of the analysis, in the form of relationships between the limiting values of utilisation time $(T_{tp})_0$ and the size of the heat and power plant at $u_{tp} = 0.5$ and at different back-pressures of steam. As can be seen from the diagrams, the limiting values $(T_{tp})_0$ decrease with the increase in output of boilers installed in heat and power plants. This means that the profitability of these heat and power plants increases, as the combined production becomes profitable starting with increasingly lower values of utilisation time.

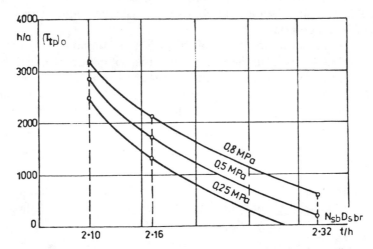

Fig. 5.10   *Profitability of a back-pressure heat and power plant depending upon the number and the rated capacity of boilers and on steam back-pressure at $u_{tp} = 0.5$*

Chapter 6

Assessment of costs in combined heat and power plants

6.1 Conditions and criteria for assessing combined production costs

The question of choosing the appropriate method of assessing the costs of the combined generation of electrical and heat energy is related to the more general problem of profitability of heat and power production. This concerns, although to a slightly different degree, both district heat and power plants and the related main heating networks, and industrial heat and power plants.

The correct method of cost assessment should be based on the following basic assumptions:

- both products of combined generation in a heat and power plant – electrical energy and heat energy – are necessary to satisfy the energy demands;
- the assessment of costs is primarily an economic and not a technical problem, hence economic criteria should have precedence over technical or physical criteria;
- a condition of combined production is the attaining of benefits on a general scale, and at the same time, the combined production should embrace all partners contributing to the increasing of these benefits;
- the correlation of benefits between both types of energy generated in a heat and power plant is proportional to the costs of separate generation.

The following requirements as to the assessment of costs in a heat and power plant supplying the main heating network ensue from the general assumptions mentioned:

1. Before proceeding to divide the costs, it should be checked whether the condition of general profitability of combined production is fulfilled

$$K^{CH} + K^{HN} < K^{CP} + K^{TL} + K^{HP} \tag{6.1}$$

where:
K^{CH} – the annual cost of producing both kinds of energy in the heat and power plant investigated;

K^{HN} – annual cost of transporting heat in the main heating network, to the points where the heat is distributed and directed to the local networks;
K^{CP} – the annual cost of generating electrical energy in a substitute condensing power plant;
K^{TL} – the annual cost of transmitting electrical energy along power lines from the power plant to the point at which the heat and power plant would be located;
K^{HP} – the annual cost of producing heat energy in a substitute heating plant supplying the local network.

2. As result of the assessment the costs apportioned to the given energy at the input to the distribution network should be lower in a combined than in a separate system. Thus in the case of electrical energy there should be the inequality

$$K_E^{CH} < K^{CP} + K^{TL} \tag{6.2}$$

and in the case of heat energy

$$K_H^{CH} + K^{HN} < K^{HP} \tag{6.3}$$

where:
K_E^{CH} – annual production cost in a heat and power plant apportioned to electrical energy;
K_H^{CH} – annual production cost in a heat and power plant apportioned to heat energy;
here

$$K_E^{CH} + K_H^{CH} = K^{CH} \tag{6.4}$$

3. In the case of approximate calculations, omitting the costs of energy transmission, the above inequalities assume a slightly simpler form

$$K^{CH} < K^{CP} + K^{HP} \tag{6.5}$$

$$K_E^{CH} < K^{CP} \tag{6.6}$$

$$K_H^{CH} < K^{HP} \tag{6.7}$$

4. The cost of heat should depend on the parameters of steam delivered from the heat and power plant in such a manner that this cost would be lower at lower parameters. This would enable the sale of heat at a lower price and thus encourage the lowering of pressure of steam delivered to consumers, which is beneficial from the point of view of energy.

5. The cost of heat should depend upon the utilisation time and season. The longer the utilisation time, the cheaper the cost of heat. The heat consumed in summer should be cheapest, when the heat and power plant has no heat load.

6. The cost of electrical energy should also depend upon the utilisation time, as well as the time of day and season, in accordance with the known laws of energy economy.

7. The assessment of costs should depend upon the technical progress in the substitute objects (CP + HP), and can thus be based on the comparison of costs which are up-to-date at that time, if the objects were built at the same time as the given heat and power plant. For an identical heat and power plant built several years later, other substitute objects can be taken, e.g. a power plant fitted with larger units with higher steam parameters.

8. Although the breakdown of total annual costs is more correct, the possibility of applying approximations in the form of the breakdown of fixed and variable costs for relatively small changes in utilisation time, should be afforded.

9. A set of equations is required which, being theoretically correct for an ideal, solely back-pressure heat and power plant, could also be applied to an extraction-condensing heat and power plant fitted with reducing-cooling valves or water boilers for peak-load operation.

6.2 Principal methods of assessing combined production costs

The simplest theoretical case of combined heat and power production is a so-called ideal back-pressure heat and power plant from which electrical power and energy are delivered in such a relation which corresponds to the generation in the back-pressure turbine, which means without supplying live or reduced steam. Fig. 6.1 shows such an ideal system. In this diagram, various ways of assessing costs can easily be compared. The method of comparison relates the breakdown of costs to the enthalpy of the energy carrier in the heat and power plant.

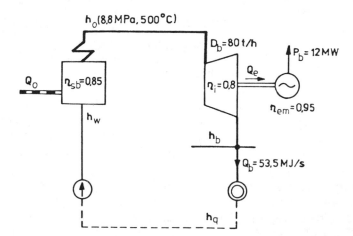

Fig. 6.1   *Diagram of a back-pressure heat and power plant to compare the methods of cost assessment*

For example, the question of the breakdown of variable costs is considered, namely, mainly the costs of fuel in combined production. It is a known fact that the combined production affords benefits, among other things, in the form of fuel savings as compared with the separate generation of electrical and heat energy. The problem does relate, in fact, to the breakdown of these advantages. Here, there are primarily two simple, but simultaneously opposing methods treating one of the forms of energy as a byproduct, which are possible:

- the physical method in which the benefits are credited to electrical energy generated in the back-pressure cycle;
- the thermodynamic method, in which almost all the benefits are credited to heat energy.

6.2.1 Physical method

The physical method consists in the division of costs being proportional to the amount of heat utilised to generate both kinds of energy in accordance with the diagram in Fig. 6.2.

Part of the costs expressed by the relationship

$$x_e = \frac{Q_e + \Delta Q_e}{Q_0} \qquad (6.8)$$

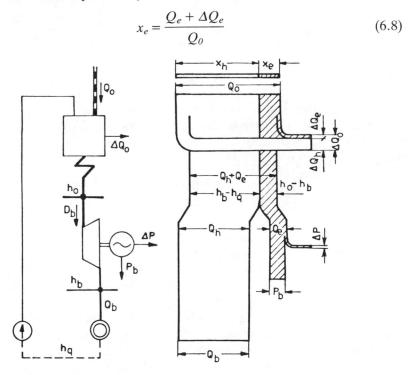

Fig. 6.2   *Diagram of cost assessment in a back-pressure heat and power plant by the physical method*

fall to electrical energy, the part which remains

$$x_h = \frac{Q_h + \Delta Q_h}{Q_0} \tag{6.9}$$

to heat energy; here the sum $\Delta Q_e + \Delta Q_h$ denotes the total heat losses in the boiler and steam pipelines, whereas the total heat delivered in the fuel

$$Q_0 = Q_e + Q_h + \Delta Q_e + \Delta Q_h \tag{6.10}$$

Assuming that the division of heat losses in the boiler and pipelines takes place in the same ratio as the division of heat delivered to the turbine set into the components Q_e and Q_h, we obtain

$$x_e = \frac{Q_e}{Q_e + Q_h} \tag{6.11}$$

$$x_h = \frac{Q_h}{Q_e + Q_h} \tag{6.12}$$

At the values of enthalpy of inlet steam h_0, outlet steam h_b and enthalpy of returning condensate h_q given in Fig. 6.1, the shares of fuel costs are

$$x_e = \frac{h_0 - h_b}{h_0 - h_q} \tag{6.13}$$

$$x_h = \frac{h_b - h_q}{h_0 - h_q} \tag{6.14}$$

At a given efficiency of steam boiler η_{sb}, efficiency of pipelines η_{pi} and electromechanical efficiency of the turbogenerator set η_{em} in the heat and power plant, the partial efficiency of generating electrical energy η_e and heat energy η_h can then be calculated

$$\eta_e = \eta_{sb}\, \eta_{pi}\, \eta_{em} \tag{6.15}$$

$$\eta_h = \eta_{sb}\, \eta_{pi} \tag{6.16}$$

As can be seen, the treating of the electrical energy generated in a combined plant as byproduct of the heat production is a characteristic feature of the physical method. The efficiency of the high-pressure steam boilers and pipelines $\eta_{sb} \cdot \eta_{pi}$ is assumed as the partial efficiency of heat production η_h, thus obtaining a very high efficiency of generating electrical energy η_e, the components of which are solely the efficiency of the boilers together with the pipelines and the electromechanical efficiency of the turbogenerator set. In view of this, the consumption and cost of fuel for electrical energy are low, as they are connected only with the drop in enthalpy in the back-pressure turbine.

The consumption and cost of fuel for heat energy, however, are high. As a result, heat delivered by the heat and power plant is expensive and together with the transmission costs may become more expensive than heat produced separately in a heating plant, although the total costs of combined production are lower than those of separate production of the two energy carriers.

The method of cost assessment based on the above-mentioned physical method has been described in detail by A. M. Komarov and V. V. Luknickij [19], as well as A. A. Lagovskij and V. B. Pakshver [30]. It is known – sometimes under other names – to almost all authors engaged in this problem, many of them agreeing that this is a theoretically inaccurate method, but it is recommended in view of simplicity of calculations irrespective of the economic consequences arising from the substantial increase in the price of heat delivered.

6.2.2 Thermodynamic method

The thermodynamic method consists in adding to the electrical power P_b actually generated in combined production, the condensing power P_c which could be obtained if the steam were expanded from the state at the back-pressure turbine outlet to the vacuum corresponding to the conditions existing in the substitute condensing power plant.

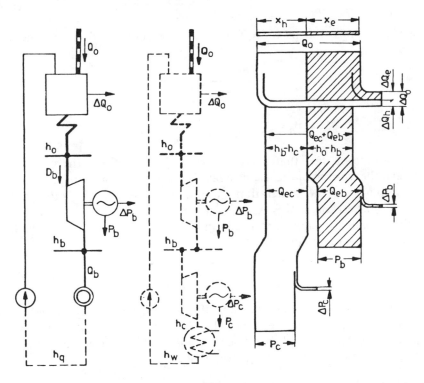

Fig. 6.3   *Diagram of cost assessment in a back-pressure heat and power plant by the thermodynamic method*

The diagram of cost assessment by the thermodynamic method is given in Fig. 6.3. In accordance with this diagram, the division of fuel costs is carried out in proportion to the consumption of heat for the production of back-pressure energy Q_{eb} and that for the previously defined condensing energy Q_{ec}.

The share of electrical energy is thus

$$x_e = \frac{Q_{eb}}{Q_{eb} + Q_{ec}} \qquad (6.17)$$

and that of heat energy is

$$x_h = \frac{Q_{ec}}{Q_{eb} + Q_{ec}} \qquad (6.18)$$

With the steam enthalpy values given for the physical method, the division of fuel costs is carried out according to the equations

$$x_e = \frac{h_0 - h_b}{h_0 - h_c} \qquad (6.19)$$

$$x_h = \frac{h_b - h_c}{h_0 - h_c} \qquad (6.20)$$

In view of this, the partial efficiencies of generating electrical and heat energy amount to

$$\eta_e = \frac{h_0 - h_c}{h_0 - h_q} \, \eta_{sb} \, \eta_{pi} \, \eta_{em} \qquad (6.21)$$

$$\eta_h = \frac{h_0 - h_c}{h_b - h_c} \cdot \frac{h_b - h_q}{h_0 - h_q} \, \eta_{sb} \, \eta_{pi} \qquad (6.22)$$

It results from this, that a characteristic feature of the thermodynamic method is the treating of the usable heat obtained in combined production as a waste product in the generation of electrical energy. The efficiency of generating electrical energy η_e is assumed to be that corresponding to a substitute condensing power plant, defined by the formula (6.21), in which the efficiency of the cycle amounts to

$$\eta_{cc} = \frac{h_0 - h_c}{h_0 - h_q} \qquad (6.23)$$

In view of this, the partial efficiency of producing heat energy η_h defined in the formula (6.22) is greater than the efficiency of the efficiency of the high-pressure boilers and pipelines $\eta_{sb} \cdot \eta_{pi}$. This is easy to prove, as $\eta_h > \eta_{sb} \, \eta_{pi}$ if the following inequality applies

$$\frac{h_0 - h_c}{h_b - h_c} \cdot \frac{h_b - h_q}{h_0 - h_q} > 1 \qquad (6.24)$$

This is true since introducing the relationship (6.23) we obtain

$$\eta_{cc}\frac{h_b - h_q}{h_b - h_c} > 1 \tag{6.25}$$

and

$$\frac{h_b - h_c}{h_b - h_q} < \eta_{cc} \text{ for all } h_b < h_0$$

When applying this method, although heat energy delivered by the heat and power plant is much cheaper, electrical energy generated in this plant may become more expensive than energy generated in modern condensing power plants, even if the total costs of combined production are lower than those of separate production of the two energy carriers.

The thermodynamic method is mainly recommended by the German authors, e.g. H. Beckmann [9] and K. Schäff [72]. This ensues, among other things, from the fact that the authors mentioned put forward the theory that the ratio of heat as a waste product to electrical energy (the main product of the heat and power plant) should not influence the shaping of prices of electrical energy, which must be set at the level of the substitute condensing power plant.

A. Fonó and F. Sóváry [11] also propose the breakdown of variable costs using the thermodynamic method. The formulae they derive for the assessment of costs lead to formulae identical with (6.19) and (6.20), with different denotations of the particular values of steam enthalpy.

6.2.3 Intermediate methods
The method of cost assessment proposed by H. Scheltz [73], is based on the assumption that the cost of each of the two kinds of energy constitutes the arithmetic mean of the costs calculated when applying the two previous methods (I and II).

Thus, for electrical energy

$$x_e = \frac{(x_e)_I + (x_e)_{II}}{2} \tag{6.26}$$

and for heat energy

$$x_h = \frac{(x_h)_I + (x_h)_{II}}{2} \tag{6.27}$$

Substituting appropriate relationships, expressed by the formulae (6.13) and (6.14), as well as (6.19) and (6.20), we obtain

$$x_e = \frac{1}{2}\left(\frac{h_0 - h_b}{h_0 - h_q} + \frac{h_0 - h_b}{h_0 - h_c}\right) \tag{6.28}$$

$$x_h = \frac{1}{2}\left(\frac{h_b - h_q}{h_0 - h_q} + \frac{h_b - h_c}{h_0 - h_c}\right) \tag{6.29}$$

Among others, A. Lévai [29] and F. Komprda [20] quote H. Scheltz's intermediate method, being aware, however, of the artificiality of the assumption (6.26), consisting in the purely mechanical choice of the 'golden mean' between two extreme methods: the physical and the thermodynamic one.

The next method of cost assessment proposed by Modrovich is quoted in F. Komprda's paper [20], as the second example of arbitrary choice of a certain intermediate state between the physical and thermodynamic methods. In this method, instead of the arithmetic mean the relationships assumed are as follows:

$$x_e = \frac{(x_e)_{II}}{1 + (x_e)_{II} - (x_e)_{I}} \tag{6.30}$$

$$x_h = 1 - x_e \tag{6.31}$$

Substituting the appropriate relationships (6.13) and (6.19) instead of $(x_e)_I$ and $(x_e)_{II}$, we obtain

$$x_e = \frac{h_0 - h_b}{h_0 - h_b + \dfrac{h_0 - h_c}{h_0 - h_q}(h_b - h_q)} \tag{6.32}$$

also from formula (6.31)

$$x_h = \frac{h_b - h_q}{h_b - h_q + \dfrac{h_0 - h_b}{h_0 - h_c}(h_0 - h_q)} \tag{6.33}$$

A. I. Andryushchenko [6] carried out the first attempt at applying such an analysis of costs which would be based on some kind of economic criterion. A. Lévai [29] and F. Komprda [20] quoted his reasoning in their papers. In this method, it is assumed that the relative fuel costs are proportional to the drops in enthalpy corresponding to the utilisation of steam in the back-pressure turbine and in the steam receiver

$$x_e = a(h_0 - h_b) \tag{6.34}$$

$$x_h = b(h_b - h_q) \tag{6.35}$$

at the same time

$$a(h_0 - h_b) + b(h_b - h_q) = 1 \tag{6.36}$$

where a and b are constants which can be defined applying the following reasoning:

– if steam with an enthalpy of h_0 were expanded in a condensing power plant and the drop in enthalpy $h_0 - h_c$ were utilised, the specific fuel cost k for the generation of this steam would amount to

$$k = a(h_0 - h_c) \qquad (6.37)$$

– if steam with an enthalpy of h_0 were supplied to the heat consumer, who would utilise the drop in enthalpy $h_0 - h_q$, the specific fuel cost for the generating of this steam would also amount to k, where

$$k = b(h_0 - h_q) \qquad (6.38)$$

By dividing both sides of the equations (6.37) and (6.38) we obtain

$$\frac{a}{b} = \frac{h_0 - h_q}{h_0 - h_c} \qquad (6.39)$$

after which the constants a and b can be found from the equations (6.36) and (6.39)

$$a = \frac{1}{h_0 - h_b + \dfrac{h_0 - h_c}{h_0 - h_q}(h_b - h_q)} \qquad (6.40)$$

$$b = \frac{1}{h_b - h_q + \dfrac{h_0 - h_b}{h_0 - h_c}(h_0 - h_q)} \qquad (6.41)$$

From this the required breakdown of costs can be found

$$x_e = \frac{h_0 - h_b}{h_0 - h_b + \dfrac{h_0 - h_c}{h_0 - h_q}(h_b - h_q)} \qquad (6.42)$$

$$x_h = \frac{h_b - h_q}{h_b - h_q + \dfrac{h_0 - h_b}{h_0 - h_c}(h_0 - h_q)} \qquad (6.43)$$

It results from the identity of formulae (6.42) and 6.43) with (6.32) and (6.33) derived from Modrovich's assumptions, that both methods lead to identical results. The same was found by F. Komprda [20] based on the conformity of numerical results obtained by the Andryushchenko's and Modrovich's methods in the numerical example he himself worked out.

The weak point of Andryushchenko's method is that he assumed an identical specific cost of producing heat in the high-pressure steam boilers of a substitute condensing power plant and in the low-pressure boilers of a substitute district heating plant.

6.2.4 Modified methods

Similar to the thermodynamic method presented previously, the exergetic method proposed by L. Nehrebecki [61] is based on conditions resulting from the principles of thermodynamics.

If the assessment of fuel costs is carried out in proportion to the exergy, which is calculated as a parameter of thermodynamic state on the basis of enthalpy h, entropy s of the given state, and the absolute ambient temperature T_{amb}, enthalpy h_{amb} and entropy s_{amb} according to the formula

$$e = h - h_{amb} - T_{amb}\,(s - s_{amb}) \tag{6.44}$$

then the corresponding formulae for the shares of electrical and heat energy are

$$x_e = \frac{e_0 - e_b}{e_0} \; ; \quad x_h = \frac{e_b}{e_0} \tag{6.45}$$

where:

e_0 – exergy of steam at the turbine inlet;

e_b – exergy of steam at the turbine outlet.

W. Sroczyński [16] proposed a simplified method of determining the value of steam in combined heat and power production for the chemical industry. This consists in the cost of outlet steam from a back-pressure turbine being calculated on the basis of the cost of live steam, which is reduced according to the value of steam enthalpy. This leads to the following division of fuel costs:

$$x_e = \frac{h_0 - h_b}{h_0} \tag{6.46}$$

$$x_h = \frac{h_b}{h_0} \tag{6.47}$$

J. Wagner [83] proposed a certain modification of the thermodynamic method as formulated by A. Fonó and E. Sóváry [11], assuming that the cost of generating electrical energy in combined heat and power production is equal to the cost of fuel in an equivalent condensing power plant built at the same time as the heat and power plant in question. The steam parameters and resulting values of heat consumption in this power plant may, however, differ from those in the heat and power plant under consideration.

In the classic thermodynamic method, identical parameters are assumed for the heat and power plant considered and for the equivalent condensing power plant, plus identical efficiency of the boilers. Given such assumptions, it can be demonstrated that the thermodynamic and Wagner's methods led to identical results.

If there are higher steam parameters and boiler efficiency in an equivalent condensing power plant, then J. Wagner's method differs from the thermo – dynamic one. The basic assumption common to both methods remains, however, that heat energy is to have the main share in the benefits from combined production, whereas electrical energy bears the costs of so or otherwise defined independent generation in an equivalent condensing power plant.

6.3 Economic method of assessing combined production costs

6.3.1 Principles of the economic method

L. Musil was the first advocate of this method. In his textbook [60], he put forward the theory that the assessment of costs in combined production should be by economic comparison with separate production. This comparison can be presented in the diagram (Fig. 6.4) in the form of a triangle.
If we denote:

K^{CH} – annual costs in the heat and power plant considered,
K^{CP} – annual costs in a substitute condensing power plant,
K^{HP} – annual costs in a substitute heating plant,

then the extreme points of the triangle are obtained on the assumption that either the electrical or the heat energy has to bear the total generating costs.

If the costs of independent generation K^{CP} and K^{HP} are denoted on the co-ordinate axes, then the straight lines MR and NR can be drawn, and the point R is determined. Connecting this point with the beginning of the system, point P is obtained at the intersection of MN with OR, this determining the division of cost K^{CH} into the components K_E^{CH} and K_H^{CH}.

The real task of L. Musil's method was to determine the ranges of variation of the specific cost of electrical energy and heat in a heat and power plant. These ranges are represented by

$$k_{E\ min} \leqslant k_E \leqslant k_{E\ max} - \text{section QM}$$
$$k_{H\ min} \leqslant k_H \leqslant k_{H\ max} - \text{section QN}$$

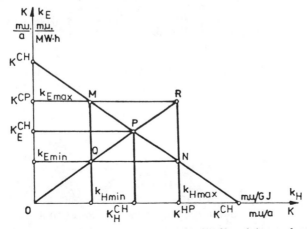

Fig. 6.4   *Comparative graph for the various methods of breakdown of costs in combined production*

These variations correspond to different positions of point P on the section MN. One of these is optimum and is determined by the straight line OR.

This method, in this general form, is unsuitable for direct application, as it does not relate the assessment of costs to the parameters of the heat and power plant. Neither are fixed and variable costs distinguished. Thus the position (angle of inclination) of the hypotenuse MN changes depending upon the utilisation time of the peak-load and the combined production index.

The method presented by the author of this book [42] was based on the criterion of the economic division of costs in an analytical formulation, approximately similar to L. Musil's graphic division.

Initially, a substitute condensing power plant was assumed to be one with the same parameters of live steam as in the heat and power plant under consideration, whereas in the substitute heating plant it was assumed that the steam pressure was equal to that of steam from the extraction (back-pressure) outlet of the given heat and power plant. The division was carried out separately for fixed and variable costs. This method could be applied for both back-pressure heat and power plants, and for extraction-condensing plants with a given index of the share of back-pressure energy.

In the author's later publications [40], [41], [48], this method has been so generalised as to enable, for comparison, to assume a substitute condensing power plant with any parameters of live steam, also those higher than in the case of the heat and power plant under consideration. In this method, division equations are obtained by reasoning based on the comparison of combined and separate production.

If combined production is more profitable, then in accordance with the inequality (6.5) there exists a positive difference between the sum of annual separate production costs in a condensing power plant (CP) and a heating plant (HP) and the annual cost of combined production in a heat and power plant (CH)

$$\Delta K = K^{CP} + K^{HP} - K^{CH} > 0 \qquad (6.48)$$

The division of cost K^{CH} into components corresponding to electrical energy K_E^{CH} and heat energy K_H^{CH} is reduced to the division of the difference of costs ΔK into parts ΔK_E and ΔK_H, where

$$\Delta K_E = K^{CP} - K_E^{CH} \qquad (6.49)$$

$$\Delta K_H = K^{HP} - K_H^{CH} \qquad (6.50)$$

The missing equations are obtained from the economic criterion of the division of combined production costs, which assumes that savings ΔK_E and ΔK_H, falling to both kinds of energy, should be related in the same ratio as the costs of separate generation of energy

$$\frac{\Delta K_E}{\Delta K_H} = \frac{K^{CP}}{K^{HP}} \tag{6.51}$$

The required division of total annual costs of a heat and power plant is determined from the four equations presented

$$x_e = \frac{K_E^{CH}}{K^{CH}} = \frac{K^{CP}}{K^{CP} + K^{HP}} \tag{6.52}$$

$$x_h = \frac{K_H^{CH}}{K^{CH}} = \frac{K^{HP}}{K^{CP} + K^{HP}} \tag{6.53}$$

In view of this, the annual costs falling to both types of energy amount to

$$K_E^{CH} = K^{CH} \frac{K^{CP}}{K^{CP} + K^{HP}} \tag{6.54}$$

$$K_H^{CH} = K^{CH} \frac{K^{HP}}{K^{CP} + K^{HP}} \tag{6.55}$$

where:

K^{CH} – the total annual costs of a heat and power plant;

K_E^{CH} – the annual costs of a heat and power plant apportioned to electrical energy;

K_H^{CH} – the annual costs of a heat and power plant apportioned to heat energy;

K^{CP} – the total costs of generating electrical energy in a substitute condensing power plant;

K^{HP} – the total annual costs of generating heat energy in a substitute heating plant.

Given such a division of total annual costs, the specific costs of generating electrical and heat energy in a heat and power plant amount, correspondingly, to

$$k_E^{CH} = \frac{K_E^{CH}}{E_{CH}} \tag{6.56}$$

$$k_H^{CH} = \frac{K_H^{CH}}{H_{CH}} \tag{6.57}$$

where:

k_E^{CH} – the specific cost of electrical energy in a heat a power plant, m.u./MWh*;

k_H^{CH} – the specific cost of heat energy in a heat and power plant, m.u./GJ;

E_{CH} – the annual production of electrical energy in a heat and power plant, MWh/a;

* m.u. = monetary unit

H_{CH} – the annual production of heat energy in a heat and power plant, GJ/a.

If the difference in annual costs in favour of combined production amounts to

$$\Delta K = K^{CP} + K^{HP} - K^{CH} \tag{6.58}$$

then the specific costs of generating both types of energy in a heat and power plant can also be determined from the following relationships

$$k_E^{CH} = (1 - \delta)\, k^{CP} \tag{6.59}$$

$$k_H^{CH} = (1 - \delta)\, k^{HP} \tag{6.60}$$

where:

k^{CP} – the specific cost of generating electrical energy in a substitute condensing power plant, m.u./MWh;

k^{HP} – the specific cost of producing heat energy in a substitute heating plant, m.u./GJ;

δ – the relative difference in annual costs,
where:

$$\delta = \frac{\Delta K}{K^{CP} + K^{HP}} \tag{6.61}$$

The assessment of costs of combined energy production can be carried out separately for fixed and variable costs in a heat and power plant. For fixed costs, the relationship obtained is then

$$K_{cE}^{CH} = K_c^{CH} \frac{K_c^{CP}}{K_c^{CP} + K_c^{HP}} \tag{6.62}$$

$$K_{cH}^{CH} = K_c^{CH} \frac{K_c^{HP}}{K_c^{CP} + K_c^{HP}} \tag{6.63}$$

in which:

K_c^{CH} – the annual fixed costs of a heat and power plant;

K_{cE}^{CH} – the annual fixed costs of a heat and power plant apportioned to electrical energy;

K_{cH}^{CH} – the annual fixed costs of a heat and power plant apportioned to heat energy;

K_c^{CP} – the annual fixed costs of generating electrical energy in a substitute condensing power plant;

K_c^{HP} – the annual fixed costs of producing heat energy in a substitute heating plant, m.u./a.

With such a division of annual fixed costs, the specific fixed costs related to the attainable electrical power and heat output in a heat and power plant amount, correspondingly, to

$$k_{cE}^{CH} = \frac{K_{cE}^{CH}}{P_{CH}} \qquad (6.64)$$

$$k_{cH}^{CH} = \frac{K_{cH}^{CH}}{Q_{CH}} \qquad (6.65)$$

where:

k_{cE}^{CH} – the specific fixed annual cost of a heat and power plant apportioned to the attainable electrical power, m.u./(MW · a);

k_{cH}^{CH} – the specific fixed annual cost of a heat and power plant apportioned to the attainable heat output, m.u./(MJ · s^{-1} · a);

P_{CH} – the attainable electrical power of a heat and power plant, MW;

Q_{CH} – the attainable heat output in a heat and power plant, MJ/s.

Analogically, the division of variable costs of combined energy production can be carried out in the following way:

$$K_{vE}^{CH} = K_{v}^{CH} \frac{K_{v}^{CP}}{K_{v}^{CP} + K_{v}^{HP}} \qquad (6.66)$$

$$K_{vH}^{CH} = K_{v}^{CH} \frac{K_{v}^{HP}}{K_{v}^{CP} + K_{v}^{HP}} \qquad (6.67)$$

where:

K_{v}^{CH} – the annual variable costs of a heat and power plant;

K_{vE}^{CH} – the annual variable costs of a heat and power plant apportioned to electrical energy;

K_{vH}^{CH} – the annual variable costs of a heat and power plant apportioned to heat energy;

K_{v}^{CP} – the annual variable costs of generating electrical energy in a substitute condensing power plant;

K_{v}^{HP} – the annual variable costs of producing heat energy in a substitute heating plant, m.u./a.

With such a division of annual variable costs, the specific variable costs of generating electrical and heat energy in a heat and power plant amount, correspondingly, to

$$k_{vE}^{CH} = \frac{K_{vE}^{CH}}{E_{CH}} \qquad (6.68)$$

$$k_{vH}^{CH} = \frac{K_{vH}^{CH}}{H_{CH}} \qquad (6.69)$$

where:

k_{vE}^{CH} – the specific variable annual cost of generating electrical energy in a heat and power plant, m.u./MWh;

k_{vH}^{CH} – the specific variable annual cost of producing heat energy in a heat and power plant, m.u./GJ.

Introducing by analogy, to the formula (6.58), the difference of fixed and variable costs

$$\Delta K_c = K_c^{CP} + K_c^{HP} - K_c^{CH} \tag{6.70}$$

$$\Delta K_v = K_v^{CP} + K_v^{HP} - K_v^{CH} \tag{6.71}$$

and the coefficients

$$\delta_c = \frac{\Delta K_c}{K_c^{CP} + K_c^{HP}} \tag{6.72}$$

$$\delta_v = \frac{\Delta K_v}{K_v^{CP} + K_v^{HP}} \tag{6.73}$$

one can determine separately the specific fixed and variable costs for both types of energy as in the formulae (6.59) and (6.60)

$$k_{cE}^{CH} = (1 - \delta_c)\, k_c^{CP} \tag{6.74}$$

$$k_{cH}^{CH} = (1 - \delta_c)\, k_c^{HP} \tag{6.75}$$

$$k_{vE}^{CH} = (1 - \delta_v)\, k_v^{CP} \tag{6.76}$$

$$k_{vH}^{CH} = (1 - \delta_v)\, k_v^{HP} \tag{6.77}$$

where:

k_c^{CP} – the specific fixed annual cost of the equivalent electrical power in a substitute condensing power plant, m.u./(MW $\cdot$ a);

k_c^{HP} – the specific fixed annual cost of the equivalent heat output in a substitute heating plant, m.u./(MJ $\cdot$ s^{-1} $\cdot$ a);

k_v^{CP} – the specific variable annual cost of electrical energy in a substitute condensing power plant, m.u./MWh;

k_v^{HP} – the specific variable annual cost of heat energy in a substitute heating plant, m.u./GJ.

Attention should, however, be paid to the fact that although the difference of total annual costs ΔK, defined in the formula (6.58) is positive, which constitutes a condition for the profitability of a heat and power plant, the difference in fixed annual costs ΔK_c, defined in the formula (6.70), may be negative, as the investment outlays for the construction of a heat and power plant *(CH)* may be greater than the equivalent investment outlays in a substitute condensing power plant and heating plant *(CP + HP)*. In view of this, the relative difference δ_c in the formulae (6.72), (6.74), (6.75) may also be negative. On the other hand, the relative difference of variable costs δ_v in the formulae (6.73), (6.76) and (6.77) is always positive.

Knowing the specific fixed and variable costs of electrical and heat energy in a combined system, calculated from the formulae (6.74–6.77), one can easily make the combined specific costs of both types of energy subject to the utilisation times of the corresponding electrical and heat output attainable. The relationships obtained are

$$k_E^{CH} = \frac{k_{cE}^{CH}}{T_e^{CH}} + k_{vE}^{CH} \tag{6.78}$$

$$k_H^{CH} = \frac{k_{cH}^{CH}}{T_h^{CH}} + k_{vH}^{CH} \tag{6.79}$$

in which:

$T_e^{CH} = \dfrac{E_{CH}}{P_{CH}}$ – the annual utilisation time of the attainable electrical power in a heat and power plant, h/a;

$T_h^{CH} = \dfrac{H_{CH}}{Q_{CH}}$ – the annual utilisation time of the attainable heat output in a heat and power plant, s/a.

6.3.2 Assessment of costs in a back-pressure heat and power plant

In applying the method of the economic division of costs in combined production, the fixed and variable costs of a heat and power plant can be divided separately. It is here assumed that in a substitute condensing power plant, the steam parameters are different and usually higher than in the heat and power plant under consideration. The specific consumption of heat to generate electrical energy in a condensing turbine q_{TCP} thus takes into account the possible effect of resuperheating of steam and a complex feed-water heating system.

Assuming the denotations of steam enthalpy as given in Fig. 6.1, the variable costs per hour are obtained:

in a substitute condensing power plant

$$K_v^{CP} = \frac{c_f q_{TCP}}{\eta_{bCP}} D_b(h_0 - h_b)\, \eta_{em} \tag{6.80}$$

in a substitute heating plant

$$K_v^{HP} = \frac{c_f}{\eta_{bHP}}\, D_b(h_b - h_q) \tag{6.81}$$

where:

c_f – price of heat energy contained in the fuel, which, in this case, is assumed to be the same for a condensing power plant and heating plant;

η_b – boiler efficiency, correspondingly in a condensing power plant (CP) and heating plant (HP);

D_b – steam flow delivered from a back-pressure heat and power plant;
η_{em} – electromechanical efficiency of a back-pressure turbogenerator set.

After conversion of the formulae (6.66) and (6.67), the following division of variable costs in a heat and power plant is obtained:

$$x_{ve} = \frac{h_0 - h_b}{h_0 - h_b + \varphi_v(h_b - h_q)} \tag{6.82}$$

$$x_{vh} = \frac{\varphi_v(h_b - h_q)}{h_0 - h_b + \varphi_v(h_b - h_q)} \tag{6.83}$$

$$\varphi_v = \frac{1}{\eta_{em}\, q_{TCP}} \cdot \frac{\eta_{bCP}}{\eta_{bHP}} \tag{6.84}$$

In the particularly simple case which results from the assumption that the parameters of live steam in a substitute condensing power plant are equal to those of the heat and power plant considered, omitting the resuperheating and feed-water heating, the following relationship occurs

$$q_{TCP} = \frac{h_0 - h_w}{h_0 - h_c} \cdot \frac{1}{\eta_{em}} \tag{6.85}$$

Assuming equal values of the feed-water enthalpy

$$h_{wCP} = h_{wCH} = h_q \tag{6.86}$$

we obtain

$$\varphi_v = \frac{h_0 - h_c}{h_0 - h_q} \cdot \frac{\eta_{bCP}}{\eta_{bHP}} \tag{6.87}$$

thus

$$x_{ve} = \frac{h_0 - h_b}{h_0 - h_b + \dfrac{h_0 - h_c}{h_0 - h_q}(h_b - h_q)\dfrac{\eta_{bCP}}{\eta_{bHP}}} \tag{6.88}$$

$$x_{vh} = 1 - x_{ve} \tag{6.89}$$

Equations (6.88) and (6.89) would be identical with those given in the Andryushchenko-Komprda-Modrovich method, if the boiler efficiencies η_{bCP} and η_{bHP} were identical. Assuming the coefficient of division of costs to be φ_v according to (6.84), equations (6.82) and (6.83) are, however, a more general form of solving the question of dividing fuel costs in a back-pressure heat and power plant, as compared with the equations given by F. Komprda [20] and those determined by the author of this book in [42], as they enable the introduction, for comparison, of condensing power plants with optional parameters, for which the specific consumption of heat by the turbine q_{TCP} and the efficiency of the boiler together with pipelines η_{bCP} are known.

The division of fixed annual costs of a back-pressure heat and power plant is considered as is that of variable annual costs of the same plant. Denoting the specific fixed annual costs per unit of generating capacity in a substitute condensing plant as k_c^{CP} and corresponding fixed annual costs per unit of heat output in a heating plant as k_c^{HP} we obtain, in accordance with the formulae (6.62) and (6.63), the division of fixed costs according to the relationships which, after substitutions and transformations assume their final form, analogical to the equations (6.82) and (6.83), previously derived for the variable costs

$$x_{ce} = \frac{h_0 - h_b}{h_0 - h_b + \varphi_c(h_b - h_q)} \tag{6.90}$$

$$x_{ch} = \frac{\varphi_c(h_b - h_q)}{h_0 - h_b + \varphi_c(h_b - h_q)} \tag{6.91}$$

where

$$\varphi_c = \frac{1}{\eta_{em}} \frac{k_c^{HP}}{k_c^{CP}} \tag{6.92}$$

The difference between the method proposed here and the division of fixed costs using the Andryushchenko-Komprda-Modrovich method results from the formula (6.92) on the coefficient of division of fixed costs φ_c. The equivalence of both methods would occur in the case of the equality

$$\varphi_c = \frac{h_0 - h_c}{h_0 - h_q} = \frac{1}{\eta_{em}} \frac{k_c^{HP}}{k_c^{CP}} \tag{6.93}$$

which, of course, with the exception of a specific case is not fulfilled as the values of k_c^{HP} and k_c^{CP} depend, among other things, on the investment outlays, whereas the values of enthalpy depend on the parameters of energy carrier.

The formulae (6.90) and (6.91) constitute more general solutions of the division of fixed costs than given by the author of this book in [42], as they enable the introduction, for comparison, of condensing power plants with optional parameters, for which the specific fixed costs k_c^{CP} are known.

Example 6.1
Determine the division of costs in a 12 MW back-pressure heat and power plant which delivers all the outlet steam to heat receivers. The heat diagram of

the heat and power plant is presented in Fig. 6.1. The following fixed parameters and efficiencies of equipment are assumed:

steam pressure at turbine inlet	$p_0 = 8 \cdot 8$ MPa
steam temperature at turbine inlet	$t_0 = 500$ °C
back-pressure of steam at turbine outlet	$p_b = 0 \cdot 5$ MPa
temperature of return condensate (100% return)	$t_q = 100$ °C
temperature of feed water	$t_w = 100$ °C
ambient temperature	$t_{amb} = 15$ °C
pressure in substitute power plant condenser	$p_c = 5$ kPa
internal efficiency of turbine	$\eta_i = 0 \cdot 80$
mechanical efficiency of turbogenerator set	$\eta_m = 0 \cdot 98$
electrical efficiency of generator	$\eta_g = 0 \cdot 97$
efficiency of high-pressure boilers with steam pipelines	$\eta_{bCH} = \eta_{bCP} = 0 \cdot 85$
efficiency of boilers in substitute heating plant	$\eta_{bHP} = 0 \cdot 60$
specific heat demand of turbines in a substitute condensing power plant	$q_{TCP} = 9 \cdot 21$ MJ/kWh
unit cost of fuel	$c_f = 700$ m.u./t*
	$= 23 \cdot 9$ m.u./GJ
peak-load utilisation time	$T_{bp} = 5000$ h/a

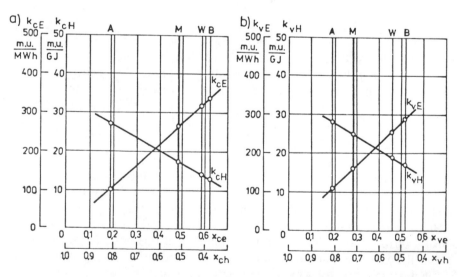

Fig. 6.5   *Results of calculations of specific fixed and variable costs of electrical and heat energy in a back-pressure heat and power plant, using various methods of assessing costs: (a) specific fixed costs, (b) specific variable costs*

* m.u. = monetary unit

and the following values of specific investment costs:

in a back-pressure heat and power plant $k_i^{CH} = 15\ 000$ m.u./kW
in a substitute condensing power plant $k_i^{CP} = 10\ 500$ m.u./kW
in a substitute heating plant $k_i^{HP} = 1500$ m.u./(kJ·s^{-1})

The results of calculations of fixed and variable costs of electrical and heat energy are presented in Fig. 6.5. The relative costs defined by the formulae (6.82) and (6.83) also (6.90) and (6.91) are denoted on the axis of abscissae, with the specific annual fixed and variable costs – on the axis of ordinates. The results of calculations of total specific annual costs in the same heat and power plant are presented in Fig. 6.6. The following denotations of the specific methods have been assumed:

A – physical method;
B – thermodynamic method;
W – Wagner's method;
M – Musil-Marecki's method of economic division of costs (thick line).

Table 6.1 gives a list of numerical results of calculations.

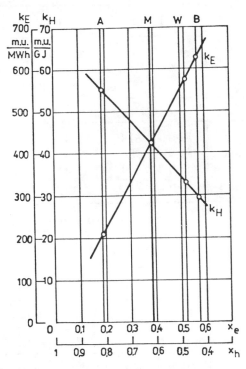

Fig. 6.6   *Results of global calculations of specific annual costs of electrical and heat energy in an example of a back-pressure heat and power plant, using various cost assessment techniques*

Table 6.1 *Assessment of costs in a back-pressure heat and power plant – comparison of results*

Method of division	Fixed annual costs					
	relative		total		specific	
	x_{ce}	x_{ch}	K_{cE}	K_{cH}	k_{cE}	k_{cH}
	%	%	$\dfrac{10^6\,\text{m.u.}}{\text{a}}$	$\dfrac{10^6\,\text{m.u.}}{\text{a}}$	$\dfrac{\text{m.u.}}{\text{MWh}}$	$\dfrac{\text{m.u.}}{\text{GJ}}$
A physical	19·0	81·0	6·2	26·3	103	27·2
B thermodynamic	61·8	38·2	20·1	12·4	335	12·9
W Wagner's	58·4	41·5	19·0	13·5	317	14·1
M Musil-Marecki's	48·5	51·5	15·8	16·7	263	17·4

	Variable annual costs					
	relative		total		specific	
	x_{vc}	x_{vh}	K_{vE}	K_{vH}	k_{vE}	k_{vH}
	%	%	$\dfrac{10^6\,\text{m.u.}}{\text{a}}$	$\dfrac{10^6\,\text{m.u.}}{\text{a}}$	$\dfrac{\text{m.u.}}{\text{MWh}}$	$\dfrac{\text{m.u.}}{\text{GJ}}$
A physical	19·0	81·0	6·4	27·1	107	28·2
B thermodynamic	52·0	48·0	17·4	16·1	290	16·7
W Wagner's	46·0	54·0	15·4	18·1	257	18·9
M Musil-Marecki's	28·6	71·4	9·6	23·9	160	24·9

	Total annual costs					
	relative		total		specific	
	x_e	x_h	K_E	K_H	k_E	k_H
	%	%	$\dfrac{10^6\,\text{m.u.}}{\text{a}}$	$\dfrac{10^6\,\text{m.u.}}{\text{a}}$	$\dfrac{\text{m.u.}}{\text{MWh}}$	$\dfrac{\text{m.u.}}{\text{GJ}}$
A physical	19·0	81·0	12·6	53·4	210	55·4
B thermodynamic	56·8	43·2	37·5	28·5	625	29·6
W Wagner's	52·2	47·8	34·4	31·6	574	33·0
M Musil-Marecki's	38·5	61·5	25·4	40·6	423	42·3

6.3.3 Assessment of costs in an extraction-condensing heat and power plant

Fig. 6.7a presents a simplified heat diagram for a heat and power plant with an extraction-condensing turbine. The electrical power generated in this system consists of back-pressure power P_b and condensing power P_c. Introducing the denotation of enthalpy h and steam flow D, as in the diagram, the following relationships for electrical power are obtained

$$P = P_b + P_c \tag{6.94}$$

$$P_b = D_b(h_0 - h_b)\,\eta_{em} \tag{6.95}$$

$$P_c = D_c(h_0 - h_c)\,\eta_{em} \tag{6.96}$$

The heat output supplied by a heat and power plant consists of heat output delivered in the back-pressure (extraction) outlet steam Q_b and heat Q_r received from live steam through the reducing valve, or produced in the peak load low-pressure boiler.

$$Q = Q_b + Q_r \tag{6.97}$$

$$Q_b = D_b(h_b - h_q) \tag{6.98}$$

$$Q_r = D_r(h_0 - h_q) \tag{6.99}$$

The true diagram can thus be exchanged for a substitute diagram (Fig. 6.7b), in which a substitute back-pressure turbine serves to generate all the

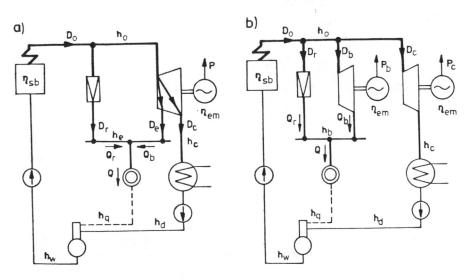

Fig. 6.7 *Diagrams of an extraction-condensing heat and power plant for comparison of cost assessment (a) diagram with extraction-condensing turbine; (b) diagram with back-pressure and condensing turbines*

back-pressure power, and a substitute condensing turbine all the condensing power.

The following denotations of the heat-power indices in a heat and power plant are assumed, as in 1.3 (Table 1.1):

$$p_a = \frac{E_b}{E} = \frac{E_b}{E_b + E_c} \quad \text{– share of back-pressure electrical energy generated,}$$

$$u_{ba} = \frac{H_b}{H} = \frac{H_b}{H_b + H_r} \quad \text{– share of back-pressure heat energy produced,}$$

$$u_{ra} = \frac{H_r}{H} = \frac{H_r}{H_b + H_r} \quad \text{– share of reducing valve in heat energy produced.}$$

The above indices with p instead of a refer to peak-load electrical power P_p and heat output Q_p.

Thus the assessment of annual variable costs should be carried out applying the following relationships in analogical form to that in equations (6.82) and (6.83):

$$x_{ve} = \frac{h_0 - h_b}{h_0 - h_b + \varphi_{ve}(h_b - h_q)} \tag{6.100}$$

$$x_{vh} = \frac{\varphi_{ve}(h_b - h_q)}{h_0 - h_b + \varphi_{ve}(h_b - h_q)} \tag{6.101}$$

but, instead of the coefficient of division of costs φ_v we insert the coefficient φ_{ve}

$$\varphi_{ve} = \frac{p_a}{u_{ba}\,\eta_{em}q_{TCP}} \frac{\eta_{bCP}}{\eta_{bHP}} \tag{6.102}$$

As can be seen, the previously introduced formula (6.84) for the system which is solely a back-pressure one, constitutes a specific form of the more general relationship (6.102), which is obtained assuming $p_a = 1$; $u_{ba} = 1$.

On carrying out the analysis of fixed costs of an extraction-condensing heat and power plant, relationships analogical to (6.90) and (6.91) are obtained, but instead of the coefficient φ_c that for the calculation of fixed costs φ_{ce} occurs

$$\varphi_{ce} = \frac{P_p}{u_{bp}\,\eta_{em}} \frac{k_c^{HP}}{k_c^{CP}} \tag{6.103}$$

which also constitutes a generalisation of the analogical formula (6.92) previously given for a system which is solely a back-pressure one.

Chapter 7

Combined gas-steam systems in heat and power plants

7.1 Review of combined gas-steam systems

Gas-steam systems consist in the combination of two thermodynamic cycles in which the working energy carrier in one is gas and in the other steam. These systems can be utilised to generate heat and electrical energy in heat and power plants. From the point of view of thermodynamic efficiency, gas-steam cycles are expedient in heat and power plants, but when assessing the profitability of their utilisation, the operational and economic conditions must also be considered.

The design of gas-steam heat and power plants so far proposed in literature and carried out, are based on greatly varied heat diagrams. The possibility of creating many different systems results from the fact that their composition includes numerous main and auxiliary appliances in both the gas and the steam cycles, and the heat carriers (steam, water, air, gas) can flow through in varied sequence. Gas-steam systems can, however, be reduced to several basic systems, if certain assumptions as to their classification are accepted.

A gas-steam system consists of four main groups of machinery and appliances:

- gas turbine set, consisting of an air compressor, gas turbine and generator, but additional appliances such as a combustion chamber, regenerative air heater, heat exchangers for interstage heating of gases and cooling of air, do not influence the classification of the systems;
- conventional steam boiler set with auxiliary appliances, the type of boiler not influencing the classification of the systems;
- steam generator set in which heat energy supplied in hot gases is used to produce superheated steam;
- steam turbine set, consisting of a steam turbine and generator together with regenerative water heaters, but the utilisation of resuperheaters and the type of turbine do not influence the classification of the systems.

The gas turbine operates in an open cycle, two systems being differentiated depending upon the direction of the gas or air flow:

– a simple system in which the working medium flows from the gas to the steam set;
– a mixed system in which, after passing from the gas to the steam part, the working medium returns to the gas part; this may take place several times.

Because of the method of utilising the working medium in the gas part, two systems are differentiated:

– one in which only the heat contained in the working medium of the gas part is utilised in the steam part;
– the other in which the heat and oxygen contained in the working medium of the gas part is utilised in the steam part.

Accepting the assumptions discussed, six basic heat diagrams of gas-steam systems in heat and power plants can be distinguished and are presented in Fig. 7.1.

Diagram 1. Liquid or gaseous fuel is supplied to the gas part, it being possible to use solid fuel also in the steam part. As compared with a steam turbine, the power rating of the gas turbine is low. This can be adopted when expanding a heat and power plant, if there exists the need for a slight increase in electrical power without the expansion of the boiler house.

Diagram 2. Liquid fuel is supplied to the gas part, and in the steam part, solid fuel can also be used. The power rating of the gas turbine is related to the flow of steam produced in the boilers in conjunction with the gas part. In steam boilers, outlet gases from the gas turbine are utilised instead of blast air. This is the most frequently adopted system, both parts of which can operate independently.

Diagram 3. In this system, only liquid fuel can be used. The steam part has a steam generator. The power ratio of the gas part to the steam part is the reverse of that in systems 1 and 2.

Diagram 4. Liquid fuel is supplied to the steam part only. A characteristic feature of this system is that the pressure in the combustion chamber is higher than atmospheric pressure. The power rating of the gas turbine is 10 ÷ 20% that of the steam turbine.

Diagram 5. This system does not require the use of liquid fuel. Heat saving is of the same order as in system 2. The characteristic feature of this system is the use of air as the working medium in the gas turbine.

Diagram 6. This system requires the use of liquid fuel only. The combustion chamber shown separately in the diagram, constitutes – as a rule – a whole with the steam generator. The opinion can be found in the literature that this system can afford the highest economic effects.

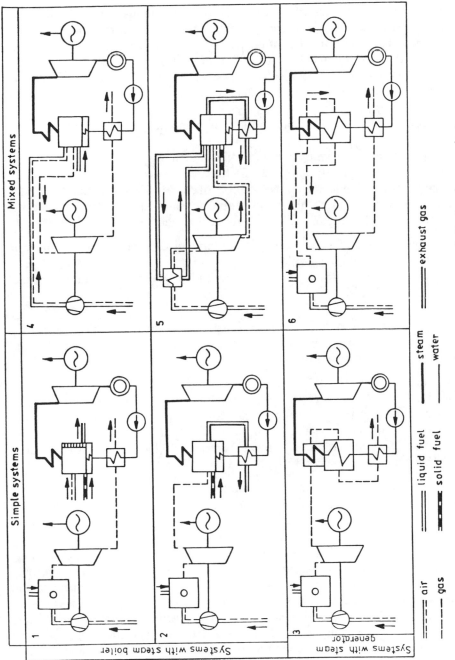

Fig. 7.1 Heat diagrams of the basic gas-steam systems in heat and power plants

7.2 Choice of gas-steam system for a heat and power plant

Each of the systems described can be employed in industrial heat and power plants with back-pressure, extraction back-pressure, or extraction-condensing turbines. The question of profitability of adopting gas-steam systems thus leads to the question of choosing the optimum of six basic systems. Within the framework of each basic system, however, there exists the possibility of various solutions of detailed heat diagrams. This section contains analysis of the influence of thermodynamic parameters on the profitability of adopting System 2 in a gas-steam heat and power plant with a back-pressure steam turbine.

Mention should be made of the fact that System 2 is at present the gas-steam system most frequently adopted. Technical and economic analyses of this system hitherto conducted on the example of condensing power plants, suggest that it may prove to be highly profitable when used in industrial heat and power plants. The arguments in favour of this include:

- an increase in the combined production index to above that attainable in steam heat and power plants;
- the fact that electrical power becomes independent, to a certain extent, of the heat loads.

Fig. 7.2 presents a diagram of an industrial heat and power plant operating in accordance with System 2. This system enables the heat and power plant to operate in the following specific cases:

- independent operation of the steam part;
- independent operation of the gas part with or without regeneration;
- combined operation of the gas and steam parts with or without regeneration in the gas part and with the separation of the flow of outlet gases from the turbine into two parts, one of which flows to the boiler as blast air and the second to the feed-water heaters.

The basic fuel in a heat and power plant is solid fuel and the use of liquid fuel can be regulated as required, depending upon the resources the plant has at its disposal. The main appliances in both gas and steam parts are typical with known and verified technical and operational indices.

The profitability of adopting gas-steam systems depends on very many parameters of different nature. Some of them (particularly those concerning the characteristics of appliances and operating conditions of the heat and power plant) do not play a significant role and the average values of such parameters found in practice can be taken in calculations. Other parameters can be taken optionally, still others result from the specific situation of the industrial plant or the economic situation of the country and have a fundamental bearing on the profitability of adopting a particular gas-steam system.

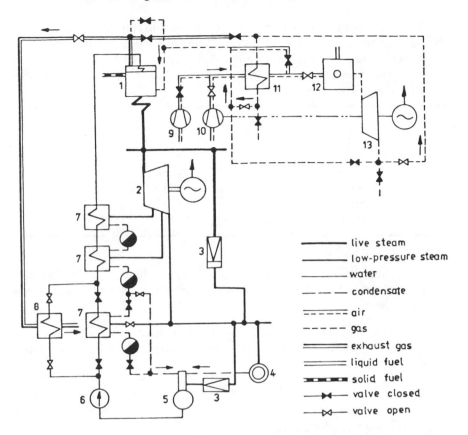

Fig. 7.2   *Heat diagram of a gas-steam heat and power plant 1 – pulverised fuel steam boiler, 2 – back-pressure steam turbine, 3 – reducing-cooling valve, 4 – heat receiver, 5 – feed-water tank with deaerator, 6 – feed pump, 7 – steam-water heater, 8 – exhaust gas-water heater, 9 – blast draught ventilator, 10 – air compressor, 11 – air heater, 12 – combustion chamber, 13 – gas turbine*

To study the question in greater detail an appropriate programme of calculations has been worked out, which will enable the determining of the thermodynamic parameters, operational conditions, and economic situation in which a gas-steam system is profitable in a back-pressure heat and power plant [27].

The parameters investigated and the initial assumptions related to these are as follows:

1. The peak-load heat output of a heat and power plant Q_p can assume arbitrary values.
2. The parameters of live steam before the steam turbine p_0, t_0 do not basically depend on the heat output. It results from previous analyses,

however, that optimum parameters of live steam can, to a certain extent, be assigned to a given heat output. The same concerns the number of stages of regenerative heating of feed water n. In view of this, these parameters are not accepted independently, but are set out in accordance with the following list:

Q_p MJ/s	p_0 MPa	t_0 °C	n —
20–60	3·4	435	1
40–120	6·4	485	2
80–240	8·8	500	4
from 160	12·5	535	4

3. The temperature of inlet gases to the gas turbine is an independent parameter and can be accepted within the limits of temperatures now controlled.

4. The peak-load heat output of the plant Q_p consists, generally, of process heat load Q_{tp} and heating load Q_{hp}, where in the case of these loads being simultaneous

$$Q_p = Q_{tp} + Q_{hp} \tag{7.1}$$

whilst the ratio of peak process to heating load is characterised by means of the share of process heat load

$$u_{tp} = \frac{Q_{tp}}{Q_p} \tag{7.2}$$

5. The annual utilisation time of process heat peak-load T_{tp} can assume arbitrary values within the limits of the number of hours in the year. For this reason, model diagrams of heat loads have been drawn, Fig. 7.3 presenting three examples of such load diagrams, namely:

$$u_{tp} = 0\cdot2 \quad T_{tp} = 4000 \ \text{h/a}$$
$$u_{tp} = 0\cdot6 \quad T_{tp} = 6000 \ \text{h/a}$$
$$u_{tp} = 0\cdot8 \quad T_{tp} = 7000 \ \text{h/a}$$

6. The heat load may be covered by the heat output delivered from the back-pressure steam turbine outlet Q_S, the share of this load in the peak-load of the heat and power plant amounting to

$$u_S = \frac{Q_S}{Q_p} \tag{7.3}$$

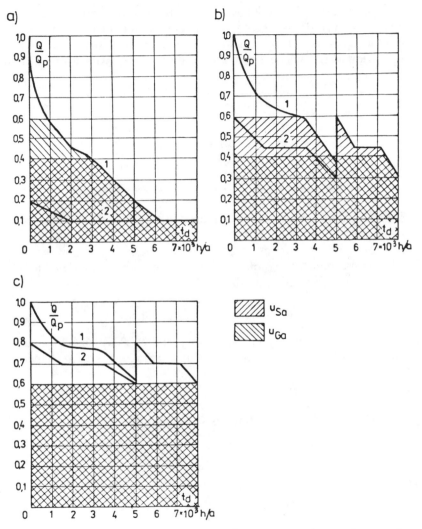

Fig. 7.3 *Models of annual heat load diagrams for a gas-steam heat and power plant: (a) $u_{tp} = 0·2$; $T_{tp} = 4000$ h/a; (b) $u_{tp} = 0·6$; $T_{tp} = 6000$ h/a; (c) $u_{tp} = 0·8$; $T_{tp} = 7000$ h/a 1 – total load, 2 – process heat load*

7. After adding the gas part to the heat and power plant, part of the heat output can be supplied with the combined operation of both parts. In view of this, the concept of the share of the gas part in covering the peak-load of the heat and power plant has been introduced

$$u_G = \frac{Q_G}{Q_p} \tag{7.4}$$

where the shares of u_S and u_G can change within the limits of 0 and 1.

8. After computing the model heat load diagrams, the annual share of heat
energy delivered from the back-pressure steam turbine outlet u_{Sa} was
obtained as was that of heat energy delivered with the combined
operation of the gas part with the steam part u_{Ga}, which was defined as

$$u_{Sa} = \frac{H_S}{H} \tag{7.5}$$

$$u_{Ga} = \frac{H_G}{H} \tag{7.6}$$

where:
H_S – the annual heat energy delivered from the outlet of a back-pressure
steam turbine
H_G – the annual heat energy delivered in co-operation with the gas and
steam parts;
H – total annual heat energy delivered by the heat and power plant.
To explain the concepts of u_{Sa} and u_{Ga}, corresponding fields for three
cases have been lined in on the load diagrams (Fig. 7.3)

$$u_S < u_G; u_S > u_G; u_S = u_G$$

Taken into account among the economic parameters which have a constant
value in a particular economic situation, but which may undergo change, are:

– index price of solid fuel c_{sf} in monetary units per mass unit of equivalent
coal;
– the ratio of the index price of liquid fuel c_{lf}, to the index price of solid fuel
c_{sf}, i.e.

$$\gamma = \frac{c_{lf}}{c_{sf}} \tag{7.7}$$

– the coefficient of annual costs r_c.

7.3 Profitability of applying gas-steam systems in heat and power plants

The programme of calculations carried out on a computer embraced the
energy balances of the individual links of heat systems in a gas, steam and
gas-steam cycle, the mass balances of all working agents participating in energy
conversions and the calculation of annual costs for the generation of electrical
and heat energy in a heat and power plant. The energy and mass balances were
carried out on the assumption that the whole stream of outlet gases from the
gas turbine flows to the boilers. In order to compare a gas-steam with a steam
heat and power plant, several technical and technico-economic indices which

might serve to estimate the profitability of a gas-steam system in a heat and power plant, were calculated. The most important of these are:

1. The net combined production index, this being understood as the ratio of the net annual production of electrical energy generated in a steam or gas-steam heat and power plant, to the annual production of heat energy with optional shares of u_S and u_G, calculated from the equation

$$\sigma_n = \frac{E_G + E_S - (\varepsilon_S - u_{Ga}\Delta\varepsilon)H}{H} \qquad (7.8)$$

where:
E_G, E_S – annual production of electrical energy in the gas and steam part;
ε_S – relative consumption of electrical energy for auxiliaries in a steam heat and power plant (related to heat energy H);
$\Delta\varepsilon$ – a decrease in the share of auxiliaries in a gas-steam plant as compared with a steam heat and power plant (related to H_G).

2. Index of the specific consumption of equivalent fuel for the generation of additional electrical energy (net), calculated as the ratio

$$b_{En} = \frac{\Delta F}{\Delta E_n} \qquad (7.9)$$

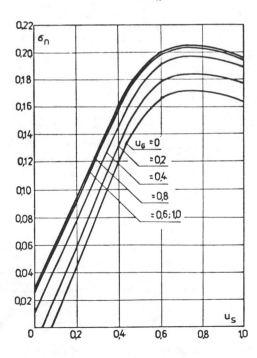

Fig. 7.4   *Index of net combined production in a gas-steam system*

where:

ΔF – annual increment of fuel consumption in a gas-steam plant as compared with a steam heat and power plant;

ΔE_n – increment of net annual production of electrical energy in a gas-steam plant as compared with a steam heat and power plant.

3. Index of the specific consumption of equivalent fuel b_b in a balancing power plant, determined from the comparison of annual costs of a gas-steam heat and power plant, related to the generation of additional electrical power and energy, with the costs of generating the same power and energy in a balancing steam power plant.

An analysis was carried out on the example of a gas-steam heat and power plant with a heat output of $Q_p = 150$ MJ/s, with the steam parameters at the steam turbine inlet of $p_0 = 8.8$ MPa, $t_0 = 500$ °C, and with the three kinds of heat load, presented in Fig. 7.3.

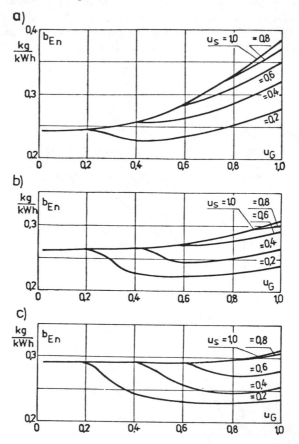

Fig. 7.5   *Specific consumption of equivalent fuel to generate additional net electrical energy in a gas-steam system: (a) $u_{tp} = 0.2$; $T_{tp} = 4000$ h/a; (b) $u_{tp} = 0.6$; $T_{tp} = 6000$ h/a; (c) $u_{tp} = 0.8$; $T_{tp} = 7000$ h/a*

Fig. 7.4 presents an example of the dependence of the net combined production index σ_n on u_S and u_G. As shown in the graph, the gas-steam heat and power plant has a higher combined production index than the steam plant. The optimum values of u_S and u_G from the point of view of maximising σ_n can be determined for each case, from such graphs.

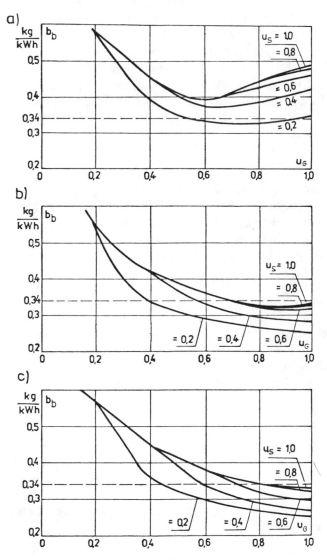

Fig. 7.6   *Limiting index of specific consumption of equivalent fuel in a balancing power plant: (a) $u_{tp} = 0 \cdot 2$; $T_{tp} = 4000$ h/a; (b) $u_{tp} = 0 \cdot 6$; $T_{tp} = 6000$ h/a; (c) $u_{tp} = 0 \cdot 8$; $T_{tp} = 7000$ h/a*

The specific fuel consumption b_{En} for the generation of additional net electrical energy for three chosen types of gas-steam heat and power plants is presented in Fig. 7.5. The optimum shares u_S and u_G are different in this case from the case of optimum shares due to the maximum index of combined production σ_n.

The limiting index of specific consumption of equivalent fuel b_b in a balancing power plant for three chosen types of gas-steam heat and power plants is presented in Fig. 7.6. It can be determined from the graphs given in this figure, whether or not a gas-steam heat and power plant is profitable, if the specific consumption of equivalent fuel in the balancing power plant is known. As an example the line $b_b = 0.34$ kg/kWh constituting the limit above which gas-steam heat and power plants are unprofitable, is marked.

In the analysis of profitability of adopting gas-steam systems presented, the influence of various parameters has been considered comprehensively. In the true conditions of an industrial plant, however, additional factors arising from the production process often occur, which will determine the most appropriate solution.

The following conclusions can be drawn from the analysis conducted:

1. The steam boiler load in a gas-steam heat and power plant is lower than in a steam heat and power plant with the same heat output. Thanks to this, in certain cases the number of boilers installed can be cut.
2. The generating capacity and electrical power attainable in a gas-steam heat and power plant, is greater than in a steam heat and power plant.
3. Gas-steam heat and power plants may be profitable in industrial plants in which the basic heat load is the process heat load with a utilisation time of $T_{tp} > 4000$ h/a.
4. The optimum number of boilers which should be fed by outlet gases from the gas turbine depends on the extent to which the heat load diagram is filled in. When the load diagrams are substantially filled in, which can be defined as the utilisation time of the peak-load $T_{tp} > 6000$ h/a, the best economic effects are attained when all the boilers are adapted for combined operation with the gas turbine.
5. In the gas part of a gas-steam heat and power plant, the aim should be to employ high temperatures of inlet gases to the turbine.
6. The profitability of employing gas-steam systems does not, in practice, depend on the price of solid fuel. The relationship between the price of a unit of heat contained in liquid fuel to that of a unit of heat in solid fuel has, however, a basic effect on the profitability of gas-steam systems. If this ratio is greater than 1.1, then gas-steam systems are only profitable in exceptional cases.

References

1. AMINOV, R. Z. and DOLGINA, V. D. (1976): 'Metodika rasčeta optimalnogo koefficienta promyšlennoj teplofikacii', *Izvestija vuzov SSSR – Ènergetika* (11), p. 82
2. ANDRZEJEWSKI, S. (1958): 'Opłacalność wytwarzania energii elektrycznej w zakładach przemysłowych', *Przegląd Elektrotechniczny* (2/3), pp. 89–92
3. ANDRZEJEWSKI, S. (1972): 'Podstawy projektowania siłowni cieplnych', Wydawnictwa Naukowo-Techniczne
4. ANDRZEJEWSKI, S. (1976): 'Skojarzone wytwarzanie energii cieplnej i elektrycznej', *Ciepłownictwo-Ogrzewnictwo-Wentylacja* (7/8), pp. 179–182
5. ANDRZEJEWSKI, S., PORTACHA, J. et al. (1966): 'Influence of heat carrier parameters on the cost of thermal energy', in: Symposium on Problems of Optimum Economic Exploitation of Energy Supply for Heating and Air Conditioning of Large Housing Developments, UN Economic Commission for Europe, Prague, Ref. C.23
6. ANDRJUŠČENKO, A. I. (1940): 'Sovetskoe kotloturbostroenie' (Moskva)
7. ANDRJUŠČENKO, A. I. (1963): 'Termodinamičeskie rasčety optimalnych parametrov teplovych èlektrostancij' (Vysšaja Škola)
8. BACHL, H. (1961): 'Energiebilanz und Rentabilität von Heizkraftwerken' (Springer)
9. BECKMANN, H. (1953): 'Die Verteilung der Selbstkosten in Industrie- und Heizkraftwerken auf Strom und Heizdampf', *Brennstoff-Wärme-Kraft*, pp. 37–44
10. FIPACE (FÉDÉRATION INTERNATIONALE DES PRODUCTEURS AUTO-CON-SOMMATEURS INDUSTRIELS D'ÉLECTRICITÉ) (1961): 'Production combinée d'énergie électrique et de vapeur par les installations à contre-pression dans l'industrie' (Bruxelles)
11. FONÓ, A. and SÓVÁRY, E. (1957): 'Die in Rechnung zu stellenden Selbstkosten der Arbeit geleisteten Abwärme', in: Weltkraftkonferenz, Beograd, Ref. B5-12
12. FRATZSCHER, W. (1961): 'Zum Begriff des exergetischen Wirkungsgrades', *Brennstoff-Wärme-Kraft,* (11), pp. 486–493
13. GAŠPAROVIĆ, N. (1962): 'Eine neue Definition des Wirkungsgrades von Heizkraftprozessen', *Brennstoff-Wärme-Kraft* (10), pp. 473–474
14. GEISSLER, T. (1957); (1958): 'Wärme- und Krafterzeugung in öffentlichen und industriellen Heizkraftwerken', *Energie* (11), pp. 427–437; (12), pp. 486–488; (2) pp. 44–49; (4), pp. 137–144
15. GOLARZ, T. (1977): 'Dobór kotłów i turbozespołów do pokrywania potrzeb technologicznych dużych zakładów przemysłowych', *Energetyka* (6), pp. 237–239
16. GÓRA, S., KOPECKI, K., MARECKI, J. and POCHYLUK, R. (1976): 'Zbiór zadań z gospodarki elektroenergetycznej', (Państwowe Wydawnictwo Naukowe)
17. HROMEK, R. (1972): 'Aufteilung der Investitions- und Erzeugungskosten in der Heizkraftwirtschaft', *Brennstoff-Wärme-Kraft*, (9)

18. Informator Energetyka (1969): (Wydawnictwa Naukowo-Techniczne)
19. KOMAROV, A. M. and ŁUKNICKIJ, V. V. (1949): 'Spravočnik dlja teplotechnikov èlektrostancij' (Gosenergoizdat)
20. KOMPRDA, F. (1948): 'Rozdělení nákladů v teplárnách, Teplárenství část I', Elektrotechnický Obzor, pp. 32–37
21. KOPECKI, K. (1976): 'Człowiek w świecie energii'. (Książka i Wiedza)
22. KOPECKI, K. (1964): 'Kryteria wyboru optymalnych nośników energii i parametrów urządzeń wytwórczych i odbiorczych w procesach technologicznych', Gospodarka Paliwami i Energią (1), pp. 17–20
23. KOPECKI, K. (1960): 'Ogólne założenia i metodyka rachunku gospodarczego w pracach planowo-projektowych w elektroenergetyce', Komitet Elektryfikacji Polski PAN, Materiały i Studia, 5 (1), (Państwowe Wydawnictwo Naukowe)
24. KOPECKI, K. (1960): 'The use of combined power and heating in industrial plants as a means of increasing energy generation efficiency in Poland', in: World Power Conference, Madrid, ref. II-A2/15
25. KOPECKI, K. (1975): 'Zagadnienia gospodarki elektroenergetycznej', Poradnik inzyniera elektryka, 4, pp. 1144–1199 (Wydawnictwa Naukowo-Techniczne)
26. KRAJEWSKI, R. (1974): 'Analiza opłacalności stosowania wybranego układu gazowoparowego w elektrociepłowniach przemysłowych', Archiwum Energetyki (1) pp. 43–62
27. KRAJEWSKI, R. and MARECKI, J. (1971): 'Economic aspects of combined gas–steam cycles applied to industrial heat and power plants', in: VIII World Energy Conference, Bucharest, Ref. 3.3–66
28. KUPPERT, H. (1969): 'Kosten und Preise für Dampf und Strom bei der Wärme-Kraft-Kupplung. Eine neue Betrachtungsweise eines alten Problems', Brennstoff-Wärme-Kraft (7), pp. 379–382
29. LÉVAI, A. (1959): 'Wärmekraftwerke, I 'Wirtschaftlichkeitsrechnung' (VEB Verlag Technik)
30. ŁAGOWSKI, A. A. and PAKSZWER, W. B. (1953): 'Elektrownie cieplne' (Wydawnictwa Naukowo-Techniczne)
31. MALEWICZ, W. (1976): 'Badanie opłacalności elektrociepłowni zawodowych', Zeszyty Naukowe Politechniki Szczecińskiej (54) pp. 1–54
32. MALEWICZ, W. (1972): 'Zastosowanie turbozespołu gazowego w elektrociepłowni miejskiej', Gospodarka Paliwami i Energią (6)
33. MANDEL, J. (1964): 'Badanie ekonomicznej efektywności inwestycji usprawniających gospodarkę energetyczną', Gospodarka Paliwami i Energią (10), pp. 325–331
34. MARECKI, J. (1956): 'Charakterystyki przenoszenia mocy dla turbozespołów w elektrociepłowniach', Przegląd Elektrotechniczny (12), pp. 499–505
35. MARECKI, J. (1961): 'Die wirtschaftlichsten Dampfzustände für Heizkraftwerke in kleinen Industrieanlagen', in: II Konferenz für Industrielle Energiewirtschaft, Budapest, Ref. E III-116
36. MARECKI, J. (1973): 'Economics of combined large-scale district heating and power production', in: First International Seminar, "Energy-73", Brussels
37. MARECKI, J. (1968): 'Lower limit of profitability of combined heat and power generation in industry', in: Symposium on the problem of electricity and heat supply for large industrial complexes, UN Economic Commission for Europe, Bucharest, Ref. B.9
38. MARECKI, J. (1965): 'Metody optymalizacji skojarzonej gospodarki energetycznej w elektrociepłowniach przemysłowych', Zeszyty Naukowe Politechniki Gdańskiej (80), pp. 3–112
39. MARECKI, J. (1969): 'Modelowe badania efektywności wytwarzania energii cieplnej w elektrociepłowniach przemysłowych i komunalnych', Zeszyty Naukowe Politechniki Gdańskiej (148), pp. 71–98
40. MARECKI, J. (1964): 'Podział kosztów w skojarzonej gospodarce cieplno-elektrycznej', Zeszyty Naukowe Politechniki Gdańskiej (44), pp. 3–52

41. MARECKI, J. (1963): 'Podział kosztów w skojarzonej gospodarce energetycznej', *Przegląd Elektrotechniczny* (1), pp. 5–11

42. MARECKI, J. (1958): 'Podział kosztów wytwarzania energii w elektrociepłowniach', *Zeszyty Naukowe Politechniki Gdańskiej* (13), pp. 79–116

43. MARECKI, J. (1977): 'Probleme der Wärme-Kraft-Kopplung im industriellen Bereich', *Energieanwendung* (10), pp. 304–308

44. MARECKI, J. (1976): 'Rola ciepłownictwa i problemy konkurencyjności z innymi sposobami zaopatrzenia w ciepło na tle światowej sytuacji paliwowo-energetycznej', *Ciepłownictwo-Ogrzewnictwo-Wentylacja* (7/8), pp. 171–175

45. MARECKI, J. (1968): 'Skojarzona gospodarka energetyczna w przemyśle (Teoria dla praktyków)', *Gospodarka Paliwami i Energią* (3); (1971), (12); (1972) (7)

46. MARECKI, J. (1964): 'The choice of an optimized investment programme for the development of combined industrial power and heat generation', in: World Power Conference, Lausanne, Ref. II B-4

47. MARECKI, J. (1968): 'The optimization of development and co-operation between combined heat and power stations and heating plants in covering the heat demand in towns', in: VII World Power Conference, Moscow, Ref. C1-52

48. MARECKI, J. (1965): 'Vorschlag zur Aufteilung der Kosten auf Elektroenergie und Wärme bei Wärme-Kraft-Kupplung', *Energietechnik* (8), pp. 361–365

49. MARECKI, J., CHERUBIN, W. and ROSADA, J. (1977): 'Stan i programy rozwoju systemów ciepłowniczych', *Postępy Techniki Jądrowej* (2), pp. 225–236

50. MARECKI, J. and GIESZCZYKIEWICZ, S. (1963): 'Ekonomiczny zasięg przesyłu ciepła przy centralizacji wytwarzania w małych elektrociepłowniach', *Gospodarka Paliwami i Energią* (8), pp. 288–292

51. MARECKI, J., KAIM, Z., SCHALLY, S. and WOJTASZEK, Z. (1969): 'Anwendungsmöglichkeiten der mathematischen Modelle für die Optimierung der Heizkraftwerke für Industriebetriebe und Gebiete in der Volksrepublik Polen', *Energietechnik* (11), pp. 484–489

52. MARECKI, J. and KRAJEWSKI, R. (1965): 'Verbunderzeugung von Kraft und Wärme in Heizkraftwerken zur Versorgung kommunaler und industrieller Abnehmer im Stadtgebiet', in: IV Konferenz für Industrielle Energiewirtschaft, Berlin, Ref. S.2–9

53. MARECKI, J. and KRAJEWSKI, R. (1963): 'Wybór parametrów pary i wielkości urządzeń energetycznych oraz ich zapotrzebowanie dla gospodarki skojarzonej w małych zakładach przemysłowych w Polsce', in: III Konferencja Energetyki Przemysłowej, Warszawa, Ref. B. 1–7

54. MARECKI, J. and SCHALLY, S. (1966): 'Economic problems concerned with covering the demand for heat in the form of domestic hot water in residential quarters of smaller towns', in: Symposium on problems of optimum economic exploitation of energy supply for heating and air conditioning of large housing developments, UN Economic Commission for Europe, Prague, Ref. C.27

55. MARECKI, J. and SCHALLY, S. (1973): 'Technisch-ökonomische Kennziffern kommunal-industrieller Heizkraftwerke und Heizwerke zur Wärmeversorgung von Städten und Industriegebieten', in: II Internationale Fernheizungs-Konferenz, Budapest, **II**, pp. 169–184

56. MARECKI, J. and SCHALLY, S. (1977): 'Optymalizacja zaopatrzenia w energię cieplną miejskich domów mieszkalnych', *Archiwum Energetyki* (2), pp. 135–150

57. MARECKI, J. and WÓJCICKI, J. (1970): 'Technical and economic aspects of district heating development in Poland', in: I International District Heating Convention, London

58. MEJRO, C. (1976): 'Ciepłownictwo a ochrona środowiska', *Ciepłownictwo-Ogrzewnictwo-Wentylacja* (7/8), pp. 175–178

59. MEJRO, C. (1974): 'Podstawy gospodarki energetycznej', (Wydawnictwa Naukowo-Techniczne)

60. MUSIL, L. (1954): 'Ogólne zasady projektowania elektrowni cieplnych' (Państwowe Wydawnictwa Techniczne)

61. NEHREBECKI, L. (1959): 'Podstawowe problemy z dziedziny układów skojarzonych w energetyce, in: Podstawowe problemy współczesnej techniki', **4**, pp. 77–122 (Państwowe Wydawnictwo Naukowe)

62. OBRĄPALSKI, J. (1955): 'Gospodarka energetyczna' (Państwowe Wydawnictwa Techniczne)

63. OCHĘDUSZKO, S. (1964): 'Termodynamika stosowana' (Wydawnictwa Naukowo-Techniczne)

64. OLSZEWSKI, A. (1974): 'Charakterystyki energetyczne turbozespołów parowych', *Prace Naukowe Instytutu Energoelektryki Politechniki Wrocławskiej* (20)

65. OLSZEWSKI, A. (1970): 'Metoda wyboru optymalnej liczby i rodzaju turbozespołów w elektrociepłowniach', (Dissertation, TU Wroclaw)

66. PIENTKA, J. (1964): 'Ocena dobroci procesów termodynamicznych i bilans strat egzergii siłowni parowej', *Zeszyty Naukowe Politechniki Poznańskiej* (24), pp. 83–119

67. PIIR, A. E. and KUNTYŠ, V. B. (1976): 'Èffektivnost' vyrabotki tepla i èlektroènergii na TEC', *Izvestija vuzov SSSR – Ènergetika* (12), p. 129

68. RAGAN, J. (1976): 'Wybór optymalnych ukladów energetycznych w elektrociepłowniach', *Ciepłownictwo-Ogrzewnictwo-Wentylacja* (7/8), pp. 201–204

69. 'Ramowe wytyczne w sprawie metodyki oceny ekonomicznej efektywności inwestycji produkcyjnych', (1974): *Monitor Polski* (28), p. 167

70. RICARD, J. (1962): 'Équipement thermique des usines génératrices d'énergie électrique', (Dunod)

71. RYŽKIN, V. Ja. (1967): 'Teplovye èlektričeskie stancii' (Ènergija)

72. SCHÄFF, K. (1955): 'Die Bewertung der Energien in Dampfkraftwerken', *Brennstoff-Wärme-Kraft* (5), pp. 202–211

73. SCHELTZ, H. (1941): 'Die Kostenanteile von Strom und Heizwärme bei Heizkraftwerken', *Archiv für Wärmewirtschaft* (12), pp. 255–258

74. SIODELSKI, A. (1971): 'Metody optymizacji skojarzonego wytwarzania ciepła i energii elektrycznej w warunkach ograniczenia łącznych nakładów inwestycyjnych', *Zeszyty Naukowe Politechniki Gdańskiej* (178), pp. 29–43

75. SIODELSKI, A. (1976): 'Programowanie rozwoju skojarzonego wytwarzania energii cieplnej i elektrycznej', *Ciepłownictwo-Ogrzewnictwo-Wentylacja* (7/8), pp. 193–196

76. SROCZYŃSKI, W. (1957): 'Obliczanie kosztów pary i energii elektrycznej w zakładach przemysłowych', *Energetyka Przemysłowa*, pp. 48–54

77. STĘPIEŃ, H. and WOJTASZEK, Z. (1976): 'Einheitliche Fernheizungsblöcke in öffentlichen Heizkraftwerken in Polen', *Fernwärme International* (2), pp. 32–36

78. SZARGUT, J. (1961): 'Bilans egzergetyczny procesów cieplnych', *Energetyka Przemysłowa* (3), pp. 73–79

79. SZARGUT, J. (1957): 'O racjonalne ustalanie cen pary', *Energetyka Przemysłowa*, pp. 104–106

80. SZARGUT, J. (1962): 'Pojęcie egzergii w odróżnieniu od energii i mozliwości praktycznego zastosowania egzergii', *Energetyka Przemysłowa* (11), pp. 374–378

81. SZUMAN, W. (1963): 'Elektrociepłownie i sieci cieplne' (Państwowe Wydawnictwo Naukowe)

82. UNICHAL (COMITÉ D'ÉTUDES DES QUESTIONS GÉNÉRALES) (1961): 'Répartition des frais dans la production combinée de chaleur et d'énergie électrique', in: Congrès de l'UNIPEDE, Baden-Baden, Ref. IIb

83. WAGNER, J. (1961): 'Zagadnienie metody podziału kosztów elektrociepłowni między oddawaną z niej energię elektryczną i parę względnie gorącą wodę', *Komitet Elektryfikacji Polski PAN, Materiały i Studia*, **5** (3), (Państwowe Wydawnictwo Naukowe)

84. WÓJCICKI, J. (1976): 'Perspektiven der Fernheizungsentwicklung bis 1990 in Polen', *Fernwärme International* (2), pp. 28–32

85. WYSOCKI, E. (1958): 'Charakterystyki energetyczne turbin parowych', *Zeszyty Naukowe Politechniki Szczecińskiej* (8), pp. 109–132

Index